The Formation of the
British State

The Formation of the British State

ENGLAND, SCOTLAND, AND THE UNION
1603–1707

Brian P. Levack

CLARENDON PRESS · OXFORD

1987

Oxford University Press, Walton Street, Oxford OX2 6DP
Oxford New York Toronto
Delhi Bombay Calcutta Madras Karachi
Petaling Jaya Singapore Hong Kong Tokyo
Nairobi Dar es Salaam Cape Town
Melbourne Auckland
and associated companies in
Beirut Berlin Ibadan Nicosia
Oxford is a trade mark of Oxford University Press

British Library Cataloguing in Publication Data
Levack, Brian P.
The formation of the British state: England,
Scotland and the union, 1603-1707.
1. Scotland—History—17th century
2. England—Foreign relations—Scotland
3. Scotland—Foreign relations—England
I. Title
327.420411 DA800
ISBN 0-19-820113-3

Library of Congress Cataloging in Publication Data
Levack, Brian P.
The formation of the British state.
Bibliography: p.
Includes index.
1. Great Britain—History—Stuarts, 1603-1714.
2. Scotland—History—Stuarts, to the Union, 1371-1707.
3. England—Foreign relations—Scotland.
4. Scotland—Foreign relations—England.
5. Scotland—History—The Union, 1707.
I. Title
DA375.L48 1987 941.06 87-12299
ISBN 0-19-820113-3

Set by Litho Link Limited, Welshpool
Printed in Great Britain
at the University Printing House, Oxford
by David Stanford
Printer to the University

Preface

THIS is a book about the union of England and Scotland from the regal union of 1603, often referred to as the Union of the Crowns, to the incorporating union or Union of the Parliaments of 1707. It is an investigation of the various proposals that Englishmen and Scotsmen made regarding the union during this period, the controversies that these proposals engendered, and the attempts that were made to implement them. It is therefore a study of the problem of Anglo-Scottish union, the problem of defining the relationship between the two countries after James VI of Scotland became James I of England in 1603. This was a matter of great importance, and at certain times, most notably in 1604, 1607, and 1707, it became the main political issue in both countries. It also proved to be an immensely difficult problem, one that challenged the minds and taxed the political abilities of both Englishmen and Scotsmen for more than one hundred years.

One of my objectives in writing this book is to reveal the enormous complexity of the union problem. The question that seventeenth- and early eighteenth-century writers and politicians asked about the union was never simply whether or not there should be a union, but what type of union, if any, there should be. The variety of union proposals that surfaced during the seventeenth and early eighteenth centuries has never been fully appreciated. These plans dealt not only with the possible union of the monarchies and parliaments of the two countries, but with the union of their central administrations, their laws, their churches, their economies, and their people. Some union projects dealt with only certain aspects of the union question, such as ecclesiastical or economic union, while others were more comprehensive in scope. As the century progressed, moreover, the terms of the union debate changed. Union did not have the same meaning in 1641 that it had had in 1604, and the union proposed in 1706 was quite different from the one that had been implemented during the Cromwellian years.

A second objective of this book is to explain why the British state which was formed in 1707 possessed the rather unusual combination of features that made it unique in Europe and qualitatively different from both the English and Scottish states out of which it was fashioned. The British state created in 1707 was by all standards an unusual structure, characterized by both unitary and pluralistic futures. It had one parliament and one central administration, but it encompassed two systems of national law and two state churches. Great Britain was also then, as it is today, a multi-national state, one in which the historic English and Scottish nations retained their own identities. A study of the union question during the 104 years that preceded the ratification of the Treaty helps to explain why the British state acquired many of these characteristics.

Although this book is very much concerned with the various union projects that were initiated between 1603 and 1707, it does not discuss them in chronological order or present a narrative account of the negotiations that attended them. In order to emphasize both the continuity and the change in men's thinking regarding the various aspects of the union question, it adopts a thematic approach. The chapters deal successively with the political, legal, religious, economic, and social aspects of the problem. This rather unconventional approach can be justified on two main grounds. First, a good amount of historical literature already exists on the various individual union projects, while William Ferguson has surveyed the entire history of English-Scottish relations before 1707.[1] Second, the different aspects of the union question were very often discussed separately in the seventeenth century, so by following a similar approach we can better understand the terms of the union debate.

In treating each aspect of the union question this book devotes an apparently disproportionate amount of attention to

[1] W. Ferguson, *Scotland's Relations with England: A Survey to 1707* (Edinburgh, 1977).

the early seventeenth century. The main reason for this is that the Jacobean union project and the large amount of writing that accompanied it have not received as much historical attention as have the projects and debates of the late seventeenth and early eighteenth centuries. It is hoped that this study will help to redress the balance. Certainly the early seventeenth-century consideration of future union, even though it achieved very little in tangible form, merits closer attention, since it established the basic terms of the union debate for the entire century. It also inspired much more thought regarding legal and ecclesiastical union than did any subsequent union project.

Throughout the book I have used the word 'state' with reference to both England and Scotland before 1707 and the United Kingdom of Great Britain after that date. The term has many different meanings, some of which may seem inappropriate when applied to Britain, which has been described as a 'stateless society'.[2] I have used the word to denote both a territorial unit that possesses sovereignty or independence from external authorities and a formal political organization whose leaders (i.e., the government) have the legally sanctioned authority to enforce obedience from the people residing within its territorial limits. The state can also be regarded, in more abstract terms, as a source of sovereignty that is above both the ruler and the ruled, but since this concept was only beginning to emerge in the seventeenth century, it will not figure in this analysis.

In quoting from original sources I have taken the liberty of modernizing spelling, capitalization, and punctuation. Titles of books, however, appear in their original form. All dates are given in Old Style, but I have considered the year to have begun on 1 January, not 25 March.

In writing this book I have incurred numerous debts, many of which I shall never be able to repay. A generous fellowship from the John Simon Guggenheim Memorial Foundation and two grants from the University Research Institute of the University of Texas at Austin enabled me to do much of the research upon which the book is based. Among the many

[2] K. Dyson, *The State Tradition in Western Europe: A Study of An Idea and an Institution* (Oxford, 1980), 19 n., 36–44.

scholars who have offered advice or criticism, given me information, or made their own work available to me are Esther
Cope, Edward Cowan, Gordon Donaldson, Elizabeth Foster,
the late Bruce Galloway, Alan Harding, Norah Fuidge, Guy
Lytle, W. R. McLeod, Hans Pawlisch, John Pocock, Wilfrid
Prest, Ian Rae, Richard Rose, Conrad Russell, David Sellar,
T. B. Smith, Nicholas Tyacke, and Philip White. My greatest
debt is to my wife, Nancy, who gave me encouragement when
the future of this project was very much in doubt and editorial
criticism when it finally began to take shape.

B.P.L.

Austin, Texas,
October 1986

Contents

List of Abbreviations

Add. MSS	British Library, Additional Manuscripts
APS	*Acts of the Parliaments of Scotland*, ed. T. Thomson and C. Innes (12 vols.; Edinburgh, 1814–44)
BL	British Library, London
Bodl.	Bodleian Library, Oxford
Bowyer Diary	*The Parliamentary Diary of Robert Bowyer, 1606–1607* ed. D. H. Willson (Minneapolis, Minn., 1931)
CJ	*The Journals of the House of Commons*, (London, 1803–)
CSPD	*Calendar of State Papers, Domestic Series* (London, 1856–72)
CSP Ven.	*Calendar of State Papers Relating to English Affairs in Venetian Archives*, ed. H. F. Brown *et al.* (London, 1864–1940)
Harl. MSS	British Library, Harleian Manuscrips
HMC	Historical Manuscripts Commission
Jacobean Union	*The Jacobean Union: Six Tracts of 1604*, ed. B. R. Galloway and B. P. Levack (Scottish History Society; Edinburgh 1985)
JR	*Juridical Review*
LJ	*The Journals of the House of Lords*, (London, 1803–)
L&L	*The Letters and Life of Francis Bacon, including all his occasional works*, ed. James Spedding (7 vols.; London, 1861–74)
NLS	National Library of Scotland
PRO	Public Record Office, London
RPCS	*Register of the Privy Council of Scotland*, ed. J. Hill and D. Masson (Edinburgh, 1877–1933)
SHR	*Scottish Historical Review*
SHS	Scottish History Society
Somers Tracts	*A Collection of Scarce and Valuable Tracts . . . Selected from . . . Libraries, particularly that of the Late Lord Somers* (13 vols.; London, 1809–15).
SP	Public Record Office, State Papers
SRO	Scottish Record Office, Edinburgh
TRHS	*Transactions of the Royal Historical Society*

I

Introduction: The Problem of Union

In March 1603 Queen Elizabeth I of England, the last of the Tudor monarchs, died at the age of sixty-nine. Since she did not have any children, and since none of her siblings survived her, the crown passed according to the English laws of inheritance to her cousin, King James VI of Scotland, who thereby became James I of England as well. This event, which marks the beginning of a new dynasty, the House of Stuart, in English history, brought about a union of England and Scotland. That union has persisted, in one form or another, until the present day.

The union of 1603, which is usually referred to as the Union of the Crowns, was very limited in scope. It was a strictly dynastic, regal, and personal union, not an incorporating union of the two kingdoms. It united the English and Scottish ruling houses and the crowns of the two countries in the person of James VI and I. It was, as James himself claimed, a union 'in his whole person', made in his blood.[1] It did not unite the laws, political institutions, or churches of the two kingdoms and did not therefore create a united kingdom, a united British state, or a single British nation. It united the kingdoms only to the extent that it gave them 'one Head or Sovereign'; it did not unite them in one body politic. James claimed that it created 'one Imperial Crown' of Great Britain, but in fact it did not even do that.[2] The crown of England remained distinct from the crown of Scotland, even though James possessed and embodied both of them. James held the two crowns in the way that an ecclesiastical pluralist held two

[1] *The Political Works of James I*, ed. C. H. McIlwain (Cambridge, Mass., 1918), 273.
[2] *Stuart Royal Proclamations*, ed. J. F. Larkin and P. L. Hughes (Oxford, 1973), i. 95; *Cobbett's Complete Collection of State Trials and Proceedings for High Treason and Other Crimes*, ed. W. Cobbett, T. B. Howell, *et al.* (London, 1809), ii. 566; 1 Jac. I, c.1.

benefices. What the regal union did, in effect, was to create a dual monarchy.

The new dual monarchy was often referred to as an empire. During the sixteenth century the word empire usually designated a sovereign territorial state which was completely independent of the pope and all foreign princes. This was clearly the meaning of the word when England was proclaimed an empire in the Act in Restraint of Appeals in 1533.[3] When applied to all of Britain, however, the word denoted an aggregation of dominions under the control of one monarch, such as the ancient Roman Empire, the medieval Angevin Empire, or the Habsburg Empire of Charles V.[4] The concept could, therefore, be extended to include Ireland and other territorial acquisitions of the crown, and in fact it soon was.[5] James was almost certainly thinking of empire in this sense when, in a reference to the Union of the Crowns, he expressed thanks that God 'has so amplified and exalted *our Empire* above all our progenitors and ancestors, so that by God's favour we have attained the monarchy of the whole of Great Britain, with all other realms and domains in highest peace by ancestral and hereditary right'.[6] The emphasis here

---

[3] 24 Henry VIII, c.12.

[4] On this shift in meaning see C. H. Firth, 'The British Empire', *SHR* 15 (1918), 185–9. The clearest illustration of the new meaning is in Ben Jonson and Thomas Dekker's poem 'Zeale', which contains the lines 'And then so rich an Empire whose fair breast / Contains four kingdoms by your entrance blest'. S. Harrison, *The Arches of Triumph Erected in Honour of the High and Mighty Prince James* (London, 1604), sig. H. The same shift can be seen in the usage of the term 'imperial monarchy'. In 1600 Thomas Wilson used this term to describe a monarchy 'held neither of Pope, Emperor nor any of but God', whereas in 1604 John Thornborough wrote that 'many shires [make] one kingdom; many kingdoms one imperial monarchy'. See T. Wilson, 'The State of England, Anno Dom. 1600'. ed. F. J. Fisher, in *The Camden Miscellany*, vol. xvi (Camden Society, 1936), 1; J. Thornborough, *The Ioiefull and Blessed Reuniting the two Mighty and Famous Kingdomes, England and Scotland* (Oxford, [1604], 7. Both meanings of empire are present in the 'imperial idea' of the 16th century. See F. Yates, *Astraea: The Imperial Theme in the Sixteenth Century* (London, 1975), 38–59. For the Jacobean exploitation of this theme and the way in which the union supported it see G. Parry, *The Golden Age Restor'd: The Culture of the Stuart Court, 1603–1642* (Manchester, 1981), 1–39.

[5] C. H. McIlwain, *The American Revolution: A Constitutional Interpretation* (New York, 1924), 26–7.

[6] W. A. Pantin, *Oxford Life in Oxford Archives* (Oxford, 1972), 49. The phrase 'Eight hundred miles his Empire goes' in a contemporary ballad suggests the new meaning. See C. H. Firth, 'Ballads Illustrating the Relations of England and Scotland during the Seventeenth Century', *SHR* 6 (1909), 114.

was upon his inheritance of a larger patrimony than any of his predecessors, the successful process of dynastic aggrandizement, not the integration of England and Scotland. The Union of the Crowns should, in fact, be considered primarily as a dynastic achievement. The ultimate product of a marriage arranged by Henry VII between his daughter Margaret and King James IV of Scotland, it succeeded mainly in expanding the dominions under the suzerainty of the House of Stuart. James achieved for the Stuart dynasty in 1603 what the members of other ruling families, such as the Habsburgs and the Valois, had sought throughout the late medieval and early modern periods—the enlargement of their territories. Marriage and the extinction of a related family line, the operating forces in the Stuart achievement, were not the only means by which such territorial aggrandizement was realized in Europe: conquest and forfeiture were just as common and sometimes even more effective. But in all cases the driving force of dynastic policy had been the desire of monarchs to enlarge their dominions, thereby increasing their wealth and military power and raising their status in the eyes of their royal confederates and rivals. The sixteenth century had been above all else an age of princes, and the rulers of both England and Scotland had pursued as avidly as any of their Continental counterparts the acquisition of territory, wealth, military power, and diplomatic prestige. The Union of the Crowns was a product of that dynastic ambition and policy, one that required as much deliberate effort, right up until 1603, as good fortune.

The flood of tracts, poems, sermons, addresses, and panegyrics that celebrated the Union of the Crowns only served to underline the regal, personal, and dynastic character of that event. Similar to, if not identical with, the congratulatory messages that greeted the king on his accession to the English throne, these encomiums were almost completely regno-centric. Indeed, they stand as some of the best examples of the cult of king-worship in the early seventeenth century.[7]

---

[7] See, for example, 'The Shepherd's Spring-song: in Gratulation of the royal, happy and flourishing Entrance, to the Majesty of England, by the most potent and prudent Sovereign, James, King of England, Scotland, France and Ireland', in *The Harleian Miscellany*, ed. T. Park (10 vols.; London, 1809–13), iii. 544; William Bellenden's verses about the union, Hatfield House, Salisbury MS 140. 102–3;

In one sermon, for example, the Scottish cleric John Gordon referred to Great Britain as the 'holy place wherein this admirable union of God and man is conjoined in the person of a Britain King'.[8] When the writers spoke about the kingdoms themselves, they referred to them in proprietary terms as 'his kingdoms', or as part of the new British 'empire'.[9] Even when they enumerated the benefits of the Union of the Crowns they underlined its regal nature, since the most obvious and immediate effects were in the area of foreign policy, which in both countries was the exclusive preserve of the king himself.[10]

While both James and his literary supporters recognized that the union of 1603 was a strictly dynastic and regal union, they also recommended in the strongest possible terms that it be perfected. Shortly after the accession of James they began to advance proposals for the integration of the two kingdoms that were now joined only in allegiance to the same king. James himself took the lead in formulating these plans for union, but he received much more political and literary support than was once commonly believed.[11] The king, who was reported 'to passionately affect the perfect uniting of both the realms',[12] proposed not only the renaming of the two kingdoms as Great Britain but the union of their laws, parliaments, councils, churches, and economies. He also called for the mutual naturalization of the subjects of both countries and the fostering of a 'union of love' between them. This was as bold and comprehensive a plan for union as anyone was to formulate during the next one hundred years. James was calling for the creation of a British national state.

M. Drayton, 'To the Majestie of King James', in *The Works of Michael Drayton*, ed. J. W. Hebel (Oxford, 1933), i. 474.

   [8] J. Gordon, *EnΩtikon, or a Sermon of the Union of Great Brittannie in Antiquitie of Language, Name, Religion and Kingdome* (London, 1604), 44.

   [9] See, for example, J. Thornborough, *A Discourse Plainely Proving the Evident Utilitie and Urgent Necessitie of the Desired Happie Union of the Two Famous Kingdomes of England and Scotland* (London, 1604), Epistle.

   [10] [Sir W. Cornwallis], *The Miraculous and Happie Union of England and Scotland* (London, 1604), sig. B4ᵛ; *The Complete Works in Verse and Prose of Samuel Daniel*, ed. A. B. Grosart (London, 1885), i. 143.

   [11] For a survey of early Jacobean literature on the union see *Jacobean Union*, pp. xxvii–xliv. For the activities of two unionists see L. L. Peck, *Northampton: Patronage and Policy at the Court of James I* (London, 1982), 186–92; K. Sharpe, *Sir Robert Cotton, 1586–1631: History and Politics in Early Modern England* (Oxford, 1979), 152–4.

   [12] SP 14/10A/17, f. 1.

There are many reasons why James and those who supported him wished to amalgamate the two kingdoms in this way. Not the least of these was to strengthen the dynastic union of 1603. James's accession to the English throne provided no guarantee that the royal succession would follow the same line in each country. As long as Scotland had its own laws of succession and its own law-making assembly, the possibility existed that the crown of Scotland could be 'alienated' from the English crown, thereby reversing the achievement of 1603. James expressed this fear as early as 1604,[13] and in fact a threat of such a Scottish defection from the English succession did eventually materialize in the early eighteenth century. At that time, as in 1604, it proved to be a main incentive for negotiating an incorporating union. Both James's union project for 1604 and the Treaty of Union approved in 1707 called for the creation of one inseparable British monarchy as well as a union of parliaments.[14]

A second reason for advocating a union of the kingdoms was the need for administrative efficiency. James in 1603 became the ruler of two separate kingdoms, each of which had its own council, parliament, and central adminstration. Since he intended to rule both countries actively, from Westminster, it was essential that a certain measure of uniformity be established between them and, if possible, that the institutions of government be actually amalgamated. James took the first steps in this process by appointing a small number of Scots to the English privy council, but an effective adminstrative union required more than that. There had to be a union of English and Scots law and a mechanism to create the same laws for both countries. This meant establishing a united state, a British body politic.

In the most general sense the union of the kingdoms promised to bring political stability to the entire island of Britain. The Union of the Crowns had virtually ended the threat of a revival of formal hostilities between the two countries, but there remained the possibility of a Scottish rebellion against the king, which did in fact occur in 1638. The

---

[13] SP 14/10A/83
[14] SP 14/9A/35; G. S. Pryde, *The Treaty of Union of Scotland and England 1707* (London, 1950), 83–4, art. 2.

integration of the adminstrative apparatus of the English and
Scottish states would prove to be an effective deterrent of such
activity, or at least it would facilitate its suppression. The same
was true for the Borders, which had a reputation for lawless-
ness. It would take more than a simple declaration that the
Borders were now the 'Middle Shires', as James proclaimed in
1603, to reduce the high level of violence and theft for which
they had become notorious.[15] What appeared to be necessary
in these circumstances was the co-ordination, if not the union,
of the judicial forces of the two states in this troubled area.

The geographical contiguity of England and Scotland
strengthened the case for a union of the two kingdoms. If the
two countries had been physically separate, a not uncommon
situation in many early modern European 'empires', then the
advantages of further union might not have been so readily
apparent. In such circumstances it might have been easier for
James to have entertained the prospect of successfully managing a dual monarchy. The geographical proximity of the two
countries, however, made the prospective incorporation of
Scotland and England into one body appear to be a natural
step, as natural as the earlier incorporation of Lancaster or
Chester into the English polity. Above and beyond these
considerations, the dominions now united under one king
constituted a natural geographical unit, and that fact
suggested that the boundaries of the territorial state should
also be those of the entire island. Unionists often used the fact
of geographical unity as an argument in favour of a further
union of the kingdoms.

Besides inhabiting the same island, the subjects of both
England and Scotland exhibited other similarities that
strengthened the commitment of James and other unionists to
the cause of closer union. Both nations were Protestant, a fact
that encouraged thoughts of ecclesiastical as well as a more
general political and social union. Their laws and political
institutions also revealed striking similarities, and they spoke
what most commentators considered to be the same language.
They conducted a limited volume of trade with each other and
shared many of the same customs and habits of dress. All

---

[15] *Stuart Royal Proclamations*, i. 18.

these similarities made it appear to James that a successful integration of the two kingdoms could be accomplished without creating a disruption of the life of either country. There remained two additional reasons why James and his supporters proposed a union of the kingdoms in the early seventeenth century. The first was that the logic of contemporary political thinking appeared to demand it. The most common way of describing the kingdom or commonwealth of England in the sixteenth and seventeenth centuries was as a body politic, of which the king was head. Now, however, as a result of the regal union, the king had 'two civil or politic bodies, his two kingdoms Scotland and England'.[16] This was at the very least an untidy situation, one which conjured up the image of a political monster, and the obvious solution to the problem was to unite the political bodies. Unionists like John Thornborough and Alberico Gentili argued quite simply that if there was a union in the head, there also had to be one in the body. The same argument appeared as late as 1689.[17]

The final reason for advocating the further union of England and Scotland was the fulfilment of a historical and divinely inspired plan. Because it was commonly believed that the two nations had been united in the murky medieval past, because many previous attempts had been made to recover that unity, especially in the sixteenth century, and because the actual Union of the Crowns appeared to be the work of Divine Providence, unionists in the early seventeenth century considered the further union of the two kingdoms to be a divine mission. The metaphysical superiority of unity over division, the unity of Christ and his church, and the more tangible benefits of political unity in other societies all tended to support this way of thinking.[18] These arguments appeared to have compelling persuasive value in the early seventeenth century. The impulse to union was as complex as it was powerful, drawing as much from abstract political and

---

[16] Harl. MS 6850, f. 63.

[17] Thornborough, *Discourse*, 17; A. Gentili, 'De Unione Regnorum Britanniae', in *Regales Disputationes Tres* (London, 1605), 44; *Vulpone: or, Remarks on Some Proceedings in Scotland relating both to the Union and Protestant Succession since the Revolution* (n.p., 1707), 14.

[18] D. J. Gordon, '*Hymenaei*: Ben Jonson's Masque of Union', in *The Renaissance Imagination*, ed. S. Orgel (Berkeley, Calif., 1975), 157–84.

theological thought as from the apparent dictates of political necessity.

James did not succeed in implementing his plan for a perfect union of the kingdoms. Opposition to virtually every aspect of his project arose either in Scotland, or in England, or in both kingdoms.[19] The king quickly retreated from his original plan, and after declaring by proclamation that he was to be styled King of Great Britain, he supported a much more limited programme for union that commissioners from both countries negotiated in 1604.[20] Even this watered-down union programme, which provided for a commercial union, an improvement of justice along the Borders, and mutual naturalization, failed to win the support of the English parliament.[21] Fearing an influx of lean Scottish kine into their green pastures, and being reluctant to give Scots trading concessions, they scrapped the entire treaty, agreeing only to repeal the hostile laws that had been passed against Scotland.

Not to be discouraged, James decided to pursue his unionist objectives through non-parliamentary means.[22] In 1608 the government instigated a legal case in which the judges determined that Scots born after the death of Elizabeth—the *post-nati*—were naturalized in England, while James used his prerogative to eliminate many restrictions on trade between the two countries. The customs on Anglo-Scottish trade were never completely lifted, but they were equalized to some extent, and Scots did not have to pay the extra alien customs on goods they imported into England.[23] At the same time James continued to pursue his goal of ecclesiastical unity by

[19] SP 14/8/9; 14/8/10; 14/28/51; M. Lee, Jr., *Government by Pen: Scotland under James VI and I* (Urbana, Ill., 1980) 34–6.

[20] For the text of the Instrument see *CJ* i. 318–23; SP 14/10B.

[21] The Scottish parliament did accept the Instrument, but their approval was contingent upon a favourable reception in England. *APS* iv. 366–71.

[22] The accusation by Dicey and Rait that James was impatient in resorting to his prerogative is unwarranted. A. V. Dicey and R. S. Rait, *Thoughts on the Union Between England and Scotland* (London, 1920), 121–2. James only used the prerogative as a last resort. See B. P. Levack, 'Toward a More Perfect Union: England, Scotland and the Constitution', in *After the Reformation: Essays on Honor of J. H. Hexter*, ed. B. Malament (Philadelphia, Pa., 1980), 60–1. For a contemporary report that the king might be 'forced' to bring about union by his own absolute power see *CSP Ven. 1603–7*, 488.

[23] S. G. E. Lythe, 'The Union of the Crowns and the Debate on Economic Integration', *Scottish Journal of Political Economy*, 5 (1958), 219–28; T. Keith, *Commercial Relations of England and Scotland, 1603–1707* (Cambridge, 1910), 16–18.

reviving the institution of episcopacy in Scotland, thus making possible uniformity and perhaps even union in church government.

Charles I never displayed his father's enthusiasm for uniting his two kingdoms, but he did continue James's efforts at establishing ecclesiastical conformity between them. Whether he planned to develop this conformity into a more general union of the kingdoms remains a matter of conjecture. In any event, the main union initiative of his reign came not from him or his council but from his ecclesiastical opponents in Scotland, the Covenanters. At the time of the Bishops' War (1640) Scottish commissioners tried to establish a religious and economic union with England.[24] This union was to have taken the form of a confederation or league, not a political union. In 1643 the Scots actually concluded an alliance with the king's English opponents in the Civil War, and as part of this alliance the subscribers pledged themselves to establish uniformity of church government. Owing to political and religious developments in England, especially the growth of anti-presbyterian sentiment, the plan did not succeed, and the Scottish Covenanters were forced to make a last desperate attempt at realizing their goal by agreeing to an Engagement with Charles I in 1647.[25] This agreement also failed to bear fruit, falling victim to the defeat of Charles and the Scots in the second English Civil War.

Although Oliver Cromwell was primarily responsible for the defeat of the Covenanters' union programme of the 1640s, he succeeded, albeit temporarily, in actually forging a union of a very different order between the two countries in the 1650s. The groundwork for this union was laid in 1651 after Cromwell's forces defeated the Scots at Worcester. At that time the English parliament appointed commissioners to present Scottish commissioners with a Tender of Union, which they accepted in 1652. The English parliament, however, failed to pass the necessary legislation to implement the union, so it was left to Cromwell to issue an ordinance to

[24] For these proposals see BL, Stowe MS 187, f. 40; NLS, Wodrow MS Fol. lxxi, no. 186.

[25] *Constitutional Documents of the Puritan Revolution, 1625–1660*, ed. S. R. Gardiner 3rd edn., (Oxford, 1906), 347–52. See also M. Lee, Jr., *The Cabal* (Urbana, Ill., 1965), 31.

this effect shortly after he became Protector. According to this arrangement, which parliament did not enact into law until 1657, Scotland sent thirty representatives to Westminster, its own parliament having been abolished.[26] In some respects the union was extraordinarily close. Complete freedom of trade was established between the two countries, and Scottish justice was administered by both English and Scottish judges. On the other hand there was no attempt to bring the churches of the two countries together, Cromwell being a firm advocate of religious toleration. There were also no attempts to establish any sort of 'union of love' between Scots and English, as King James had desired. Since Scotland was a militarily occupied country during these years, the likelihood of any such union was remote.

When the Protectorate failed in May 1659, the Cromwellian Union died with it, and although a number of attempts were made to revive it in that year, none was successful.[27] In 1660 the Restoration brought about both the removal of English troops from Scotland and a return to the old Union of the Crowns. During the following years the two countries drew not closer but further apart. Not only did their legal systems and their religious beliefs diverge considerably, but the governments of the two countries engaged in a bitter commercial rivalry. Both countries passed protective legislation against the commodities of the other, and both countries also passed navigation acts which treated the ships of the other country as alien vessels. Despite these developments, and to some extent because of the dangers that the king and his ministers detected in this rivalry, there were renewed efforts at union. The first of these, an attempt to negotiate a commercial treaty in 1668 failed, both because of English intransigence and because of an inability to reconcile the different ways in which the two governments determined economic policy.[28]

---

[26] *Acts and Ordinances of the Interregnum*, ed. C. H. Firth and R. S. Rait (London, 1911), ii. 871–5, 1131; *The Cromwellian Union: Papers Relating to the Negotiations for an Incorporating Union between England and Scotland, 1651–1652*, ed. C. S. Terry (SHS, 1902); F. D. Dow, *Cromwellian Scotland, 1651–1660* (Edinburgh, 1979).

[27] *Cromwellian Union*, pp. lxxxviii–xcvii; BL, Egerton MS 1048, ff. 176–180. For the proposal advanced by *Speculum Libertatis* see A. Woolrych, 'Last Quests for a Settlement', in *The Interregnum*, ed. G. E. Aylmer (London, 1972), 194.

[28] E. Hughes, 'Negotiations for a Commercial Union between England and Scotland in 1668', *SHR* 24 (1927), 34.

The negotiations made it clear that if any sort of union were to be agreed upon, it would have to be a parliamentary or incorporating union.[29] For this reason, and also because of a renewed desire of Charles II and his ministers to strengthen their political control of Scotland, commissioners from both countries met in 1670 to consider union. The articles of union submitted for discussion were in fact precisely the ones that James I had drafted when he first formulated a plan for union.[30] As was the case with James's plan, however, these negotiations came to naught, the principal difficulty being the English unwillingness to accept Scottish reservations regarding the proposed union of parliaments.[31]

The idea of union nevertheless survived. Scottish overtures in 1689, immediately after the Glorious Revolution, failed to elicit a positive English response, as did a somewhat feeble attempt by William III in 1700, but by 1702 the entire situation had changed. The serious possibility that Scotland would not accept the Hanoverian succession upon the death of Queen Anne and the general alienation of Scotland from England as a result of being dragged into the War of the Spanish Succession and of being discriminated against in trade threatened both the regal union itself and England's security. In such circumstances the queen and many members of the English ruling class, especially the Whigs, began openly to favour an incorporating union of the two kingdoms. At the queen's instigation commissioners from both countries were convened to discuss a union. The English Tories on the commission, however, were cool on the whole idea and negotiations eventually became deadlocked. The Scottish commissioners had agreed to accept an incorporating union in exchange for free trade with England and its colonies, but the English commissioners refused to grant Scotland sufficient compensation for the increased financial burdens of union.[32]

[29] NLS, MS 597, f. 215.

[30] Compare SP 14/9A/35 and Terry, *Cromwellian Union*, 197. For the connection between the project of James I and that of Charles II see Sir G. Mackenzie, *Memoirs of the Affairs of Scotland*, ed. T. Thomson (Edinburgh, 1821), 138.

[31] For the negotiations of 1669–70 see Lee, *The Cabal*, 43–69; W. Ferguson, *Scotland's Relations with England: A Survey to 1707* (Edinburgh, 1977), 154–7.

[32] On the union negotiations of 1689 and 1702 see Ferguson, *Scotland's Relations with England*, 170–2, 201–2; G. Omond, *The Early History of the Scottish Union Question* (Edinburgh, 1897), 147–51; T. C. Smout, 'The Road to Union', in *Britain after the Glorious*

The failure of the union negotiations of 1702–3 only aggravated the problem that had brought them about. In 1703 the Scottish parliament passed an Act of Security giving it the freedom to name its own successor to the crown upon the death of Anne, a tactic which appeared both to unionists and anti-unionists to serve their purposes.[33] The same Scottish parliament passed an Act anent Peace and War, which gave Scotland the power to accept or reject 'British' decisions to wage war or conclude peace. The English parliament responded to these measures with the so-called Alien Act of 1705, which threatened Scotland with economic sanctions and designation of all Scots not resident in England as aliens if Scotland did not either accept the Hanoverian succession or proceed with the union negotiations.[34] This Act may have marked the 'conversion' of the English parliament to union, but the successful negotiation of a treaty required less heavy-handed tactics. The return to power of the English Whigs, who were more eager for the union than the Tories, their willingness to soften the force of the Alien Act, a series of successful political manœuvers by the Court Party in Scotland, and the determination of Queen Anne to achieve union all conspired to bring about another, and this time successful, union initiative in 1706. Commissioners appointed by the queen, using the terms of the negotiations of 1702–3 as a foundation, agreed upon articles of union that were presented first to the Scottish and then to the English parliament. The Treaty encountered fierce resistance in Scotland, both in the press and in parliament, but the unionist Court Party, aided by various political pressures emanating from Westminster, prevailed. The Treaty was ratified by both parliaments and went into effect on 1 May 1707.[35]

The Treaty established a single British state—the United Kingdom of Great Britain. There was to be one imperial

Revolution, 1689–1714, ed. G. Holmes (London, 1969), 182–4; J. Mackinnon, *The Union of England and Scotland* (London, 1896), 69–75.

[33] *APS* xi. 136, c. 3.   [34] 3 & 4 Anne, c. 6.

[35] On the Treaty of 1707 see P. W. J. Riley, *The Union of England and Scotland: A Study in Anglo-Scottish Politics of the Eighteenth Century* (Manchester, 1978); Ferguson, *Scotland's Relations with England*, 232–77. The juridical bases for the implementation of the Treaty were the English Union with Scotland Act and the Scottish Union with England Act of 1707.

crown of Great Britain, one inseparable monarchy, one British parliament. The parliament was technically a new creation, distinct from both the English and Scottish parliaments that it combined, but it was in effect an enlarged English parliament to which the Scots sent forty-five MPs and sixteen elected peers.[36] In order to bring this new institution into existence Queen Anne simply proclaimed that the current English parliament, which had assembled in 1705, was now, with the addition of the Scottish members, the parliament of Great Britain.[37] The Treaty also established free trade between England and Scotland and granted Scotland the right to trade freely with the English colonies. Scotland assumed the burden of paying taxes and customs (on overseas trade) at the English rate, but in exchange for this it received the promise, which was never completely honoured, of financial compensation. The Treaty did not completely integrate the two kingdoms. Scottish judicial institutions retained their integrity, and although the Treaty made provision for the union of some of the public law of the two countries, it left the large bulk of English and Scots private law, as well as most of the criminal law of the two countries, intact. The Treaty did not even touch upon the question of ecclesiastical union, the Scottish commissioners having been forbidden to treat of such matters. Indeed, both parliaments passed acts preserving the integrity of their churches, and these two Acts of Security were appended to the Treaty itself.[38]

By formally and legally defining a new relationship between England and Scotland the Treaty of Union resolved what we may refer to as the union question or the union problem of the seventeenth century. That question was not simply whether or not there should be a further union of the two kingdoms but what type of union, if any, there should be. For more than one hundred years the rulers, statesmen, judges, members of

[36] [R. Wyllie], *A Letter concerning the Union with Sir George Mackenzie's Observations and Sir John Nisbet's Opinion upon the same Subject* (n.p., 1706), 3–4. On the British parliament's tacit assumption of the English parliament's powers see D. Daiches, *Scotland and the Union* (London, 1977), 196.

[37] Sir F. M. Powicke and E. B. Fryde (eds.), *Handbook of British Chronology*, 2nd edn. (London, 1961), 540.

[38] For the text of the Treaty, taken from the Scottish Union with England Act, see Pryde, *Treaty of Union*, 83–119.

parliament, ecclesiastical dignitaries, merchants, university officials, and to some extent the general public of both nations had dealt in one way or another with that question. It had become the subject of heated debate in the English and Scottish parliaments, and it engendered a ferocious pamphlet war in the early eighteenth century. The volume of recorded opinion on the union, in the form of speeches, letters, proclamations, and pamphlets, is truly astonishing. Between 1603 and 1707 there was no other issue in the history if either nation, with the one exception of the English Civil War, which attracted more attention and created more controversy than the union. The intensity of contemporary concern should not surprise us. The union touched on every aspect of national life and it involved a redefinition of the polity of both countries.

The union question was exceedingly complex. It involved not only a determination of the future political relationship between England and Scotland but the relationship between their systems of law, their churches, their economies, and their societies. Some of the many proposals for union that appeared between 1603 and 1707 called for union in one area of national life but not in others. In 1641, for example, Scottish commissioners sought a religious and economic but not a political union. The union project of 1668 dealt only with the economic relationship between the two countries, while the majority of plans that surfaced after 1670 envisioned a political but not a legal or religious union. To make matters even more complex, there were many different ways in which any one particular type of union could have been achieved. One could, for example, advocate the reduction of two parliaments into one or the creation of a joint parliament while retaining the original two. There could be a complete fusion of the laws of the two kingdoms or a limited union of public but not private law. There could be a union of churches according to either a presbyterian or an episcopal model, and there could be a union of doctrine and ceremonies but not of church government. The enormous range of options that contemporaries discussed and the differences that existed between the different union projects that were formulated from time to time make it difficult to label individuals as either unionists or antiunionists. Many of those who supported one type of union or one particular union project made it perfectly clear that they would not support another, while opponents of specific plans had no difficulty proclaiming their support for certain alternatives. The labels of unionist and anti-unionist have meaning only when they are used with reference to a specific type of union or a specific union project.

The chapters that follow will discuss the union question in its full complexity. They will explore the various ideas that Englishmen and Scots developed between 1603 and 1707 regarding the future relationship between the kingdoms of England and Scotland. The value of such an undertaking is twofold. On the one hand it shows how the accomplished fact of the Union of the Crowns and the perceived need for further union led men to reconsider and in some cases to change the traditional attitudes they had towards the constitution and the institutions under which they lived. The union forced men to reconsider a whole range of ideas and assumptions they had regarding the monarchy, parliament, the law, the church, the conduct of trade, and the composition of the national community. Every aspect of public life had to be rethought in terms of the union, and in the process many ideas changed.

On the other hand, the exploration of thought about the union helps to explain why the Treaty of 1707 took the form that it did and why, therefore, the United Kingdom of Great Britain acquired many of its distinctive features. This is not to suggest that the men who drafted that Treaty were engaged in some sort of philosophical enterprise. The negotiation of the Treaty was not an attempt to construct an ideal Anglo-Scottish commonwealth. The main objective of the union commissioners was to devise a union that the parliaments of both countries could accept. Since the interests of the two countries were quite different and in some cases difficult to reconcile, the Treaty necessarily involved a certain amount of compromise. It was also the work of men who were perhaps as much concerned with their political careers as they were with the interests of their countries. The making of the union has in fact been referred to as the biggest political job of the eighteenth century.[39] Nevertheless, in drafting the Treaty the commissioners had to take into consideration the attitudes

[39] W. Ferguson, 'The Making of the Treaty of Union of 1707', *SHR* 43 (1964), 110.

towards union that had developed in both countries over the previous century. These attitudes defined the limits within which compromises could be made and helped to determine many of the features of the united kingdom they constructed. The Treaty, therefore, was as much a product of the century-old union debate as it was of the immediate needs and relative bargaining strength of both parties.

Although the British polity created in 1707 did not conform to any preconceived model of the state, it did represent a modification, and in some respects a drastic modification, of a model of the state that had played a very important part in English and to a lesser extent Scottish thought regarding the union between 1603 and 1707. The model was that of the English unitary state. Since England was obviously the state with which Englishmen were most familiar, and since it was to constitute the major part of a united British kingdom, it is not surprising that many Englishmen considered the possibility of constructing a British state that possessed many of its distinctive features.

The use of the English state as a model of the future united British kingdom seemed especially appropriate to many Englishmen because England was itself in many respects a united kingdom. Not only did it represent the union of a number of once independent Anglo-Saxon kingdoms, but it comprised a number of duchies, palatinates, and principalities that had in the past possessed a certain amount of independence from the crown. All these political units had been incorporated into a single English state by the end of the sixteenth century. The last and most important of these incorporations, that of Wales, was a formal, legislatively enacted 'union', which many writers considered to be a precedent or model of the British union they were advocating or anticipating.[40]

The designation of England as a unitary state means that its various territorial components, especially those that had most recently become incorporated into it, did not possess any

For a similar opinion see B. Lenman, *The Jacobite Clans of the Great Glen, 1650–1784* (London, 1984), 74.

[40] [D. Defoe], *A Discourse upon an Union of the two Kingdoms of England and Scotland* (London, 1707), 15–16; Thornborough, *Discourse*, 9; *The Antiquity of England's Superiority over Scotland* (London, 1652), 5; BL, Stowe MS 187, f. 36.

independent legislative or judicial power. They did not possess any significant constitutional 'liberties' or privileges which exempted them from the control of the central organs of the state. They were therefore integral parts of the legal order of the entire state, upon whose sovereignty they placed no limits. A unitary state can be contrasted with what Joseph Strayer has referred to as a 'mosaic state' or a composite state, one whose constitutional structure reflects the gradual and often piecemeal incorporation of originally independent units into the body politic.[41] It can also be contrasted with a confederation, the individual components of which delegate certain powers to the central government. Finally, a unitary state differs from a federal state, the constitution of which clearly separates the powers of the central government from those of the localities.

The English state of the seventeenth century possessed a number of distinctive features which accentuated its unitary character and which, taken together, set it apart from all other European states. England was emphatically the most unitary state in seventeenth-century Europe. This is not to suggest in any way that it was the most powerful or administratively centralized. That was not the case. A relatively poor kingdom, England had neither a standing army nor a large bureaucratic establishment. Much of the administrative work of the state was conducted locally by unpaid officials, such as justices of the peace. Reliance upon local authorities does not, however, in any way detract from the unitary character of a state as long as those local authorities derive their authority from the central organs of the state and recognize their ultimate subordination to them.

The first and most important feature of the seventeenth-century English state was that its most important central authorities, the king, parliament, and the privy council, exercised effective, supreme power throughout the entire kingdom. The Tudor kings were largely responsible for this reality. Using tactics that their predecessors, the Yorkists, had developed, they succeeded in levelling the overmighty subjects of the realm and destroying the *de facto* independence of the areas these men controlled. They eliminated the liberties that

---

[41] J. R. Strayer, *On the Medieval Origins of the Modern State* (Princeton, NJ, 1970), 53.

many areas had acquired in the Middle Ages, and by virtue of the Acts of Union with Wales (1536 and 1543) shired the marcher lordships and integrated them, together with the older principality of Wales, into the English state. It is true that in both the West and the North the government continued to use special councils, the Council of the North and the Council of the Marches in Wales, which cannot be considered mere offshoots of the privy council. But the officials of these bodies received their appointments from the king and their authority from parliament, while the privy council exercised a general power of surveillance over them.[42]

The unity of the English state was most apparent in the existence of a single, sovereign parliament at Westminster. By 1604 this assembly included representatives from every English and Welsh county except Durham (which began sending MPs in 1675) as well as from a growing number of boroughs. The English parliament was an omnicompetent and sovereign legislature for the entire realm, and it did not have any competition from regional or provincial assemblies or estates. In this regard England differed most clearly from France, whose provinces retained their own estates and whose national assembly, the Estates General, gradually lost its limited powers and met for the last time before the French Revolution in 1614.

The second characteristic of the seventeenth-century English state was that it possessed a striking degree of legal unity. England had its own common law, the only distinctly national body of law in all of Europe. It acquired its common character both because the Norman and Plantagenet kings had made great efforts to enforce it throughout the entire kingdom and because the right to change that law or to make new laws was the exclusive preserve of the national parliament. When new areas were incorporated into the English state, the common law was extended to them. In Wales the replacement of native Welsh law with the common law began shortly after the English conquest of that country, but the victory of the common law was not completed until after the

---

[42] On the centralization of the Tudor state and its limitations see W. J. Jones, 'The Exchequer of Chester in the Last Years of Elizabeth I', in *Tudor Men and Institutions*, ed. A. J. Slavin (Baton Rouge, La., 1972), 125–6.

Acts of Union.[43] The judges who were appointed to administer law in the four Welsh courts of Session after the union were all trained in the common law, and they enforced English law even in the delicate area of land tenure.[44] The only area of England where the common law was not regularly administered in the late sixteenth century was in the northern Borders, where a special law, referred to either as Border or March law, was in force. After 1603, however, owing to the Union of the Crowns, Border law was abolished.[45] In one sense, therefore, the union with Scotland completed the legal unification of the southern kingdom.

The existence of a common law did not mean that England was legally monolithic. Within the various shires there existed a wide range of quite distinct local customs, all of which were easily accommodated within the common law, and there also existed outside the common law a number of special types of law, such as ecclesiastical law and the law merchant, which were administered in courts of special jurisdiction. These special laws, however, did have a standing at the common law, and their limits were defined by both parliament and the common law judges. Even more important, these laws, especially ecclesiastical law, had a national character and thus were perfectly compatible with the requirements of a unitary state. In the final analysis each English county had the same legal order, and this legal uniformity and cohesion was unmatched in any seventeenth-century European state. In France, for example, which is usually regarded as the most highly developed state in Europe in the seventeenth century, there was an astonishing degree of legal diversity, some areas being governed mainly by the civil law, others by customary feudal law.

The third characteristic of the English state was an exceptionally high degree of ecclesiastical uniformity maintained by

---

[43] R. R. Davies, 'The Twilight of Welsh Law, 1284–1536', *History*, 56 (1966), 143–64; T. A. Jones, *The Union of England and Wales* (London, 1937), 9; A. H. Dodd, ' "A Commendacion of Welshmen" ', *Bulletin of the Board of Celtic Studies*, 19 (1962), 248–9.

[44] P. R. Roberts, 'The "Acts of Union" and the Tudor Settlement of Wales', Ph.D. thesis (Cambridge, 1966).

[45] *CJ* i. 320; *RPCS* vii. 477–8; Harl. MS 292, f. 137ᵛ; *Stuart Royal Proclamations*, i. 95; SP 14/6/42.

a national or state church. Just as there was one legal order
throughout the kingdom, there was also one ecclesiastical
order. Ever since the break with Rome in 1534 that ecclesiastical order had been maintained by the organs of the state: the
king himself, who was the head of the church, and parliament,
which gradually became a partner in that headship and
passed numerous statutes enforcing religious conformity. In
England, therefore, the church became strictly subordinated
to the state and served its purposes. In such circumstances the
ecclesiastical uniformity of the country reflected the unity and
cohesion of the state itself. It is difficult to name any other
European states—except perhaps Denmark and Sweden—for
which this was true. In some Catholic countries, such as the
Spanish kingdoms and the Italian states, there was a fairly
high degree of ecclesiastical conformity, but this had much
more to do with the strength of the church than with the
power and cohesion of the state. In some Protestant states
ecclesiastical uniformity resulted from clerical control of the
state, not state control of the church. In others there was
extensive conflict between Catholics and Protestants and
hence no uniformity at all.

The fourth characteristic of the English state was the almost
complete freedom of internal trade from shire to shire. Goods
passed from one part of the country to another without paying
any customs or tariffs. Now it is true that merchants engaged
in inland or coastal trade did have to pay tolls to private lords
who had riparian rights or port fees to the towns whose
facilities they used. But they did not have to pay any fees to
cross the territorial boundaries of any one political component
of the state. England and Wales, therefore, constituted a large
free-trade area, the largest in Europe, and this situation
contributed in no mean way to the economic development of
the country. It also strengthened and underlined the unitary
character of the English state. Just as the provinces possessed
no regional assemblies, laws, or liberties, they also did not
constitute autonomous or semi-autonomous economic units.
The regulation of trade, moreover, was a strictly national,
not a regional or provincial concern.[46] It was originally the

---

[46] D. Nobbs, *England and Scotland, 1560–1707* (London, 1952), 77.

exclusive preserve of the king and his privy council, but parliament played an increasingly important role in the process during the seventeenth century.

The last salient characteristic of the English state was that it was a national state. This means that the subjects who resided within the territorial boundaries of the state considered themselves to have a distinct identity as a 'people' or 'nation'. They belonged not only to a formal political organization (the body politic) but to a community that was aware of its common language, ethnicity, history, and culture. The two types of communities are closely related, since nations, by long established definition, seek to express their identity through distinctive political institutions and because the institutions of the state often help to foster a sense of national identity. But the boundaries of states and nations do not always coincide; not all states are necessarily national states. In some cases the state is much smaller than the nation; in others it encompasses members of more than one nation. In seventeenth-century England the identity of the state and nation was, especially by current European standards, very close. England has in fact frequently been referred to as the first nation-state.[47] By the end of the sixteenth century, and indeed long before that time, the English people possessed a high degree of national consciousness—an awareness that they constituted a distinct people. This consciousness derived not only from their political, legal, ecclesiastical, and economic unity but from participation in a common culture and the use of a common language. After the defeat of the Spanish Armada this national consciousness was strengthened and heightened by the growing awareness among a large number of Englishmen that they constituted an elect nation, God's specially appointed people.[48]

The only area of the English state in which this English national consciousness was weak was in Wales, which had in

---

[47] M. Judson, *The Crisis of the Constitution* (New Brunswick, NJ, 1949), 81; J. Strayer, 'The Historical Experience of Nation Building in Europe', in *Nation-Building*, ed. K. W. Deutsch and W. J. Foltz (New York, 1963), 17–26.

[48] R. Bauckham, *Tudor Apocalypse* (Abingdon, 1978), 177–80, 235. The idea of England as the elect nation was not fully developed until the 17th century. See K. R. Firth, *The Apocalyptic Tradition in Reformation Britain, 1530–1645* (Oxford, 1979), 106–9, 167, 236.

the past been a nation itself. During the sixteenth and seventeenth centuries, however, a slow process of assimilating the Welsh, a Celtic people, to the predominantly Anglo-Saxon and Norman culture of England took place. The incorporation of Wales into the English state naturally facilitated this process. It was never completely successful, as a similar process had been earlier in Cornwall. Many Welshmen continued to speak their native Celtic language, and since the Welsh had distinctive judicial institutions there was a limited foundation for Welsh national consciousness. Nevertheless, many Welsh people, especially the gentry who were educated at the English universities, joined the English professions, and served in the English government and church, acquired a sense of identification as Englishmen. In the early seventeenth-century literature on Anglo-Scottish union the Welsh were frequently referred to as having become one people with the English.[49]

By the beginning of the seventeenth century, therefore, England was, at least by contemporary European standards, a very well-integrated state.[50] This is not to suggest that it was some sort of monolith. England had the common law, but it did not possess a uniform, hierarchically organized system of courts, as mid-seventeenth-century law reformers were quick to point out.[51] It had a state church, but there were many Catholics and Protestant nonconformists who refused to accept its authority. It had free trade from shire to shire, but a national economy was only in the early stages of formation.[52] And although many Englishmen were conscious of their national identity, their outlook was still predominantly provincial or local. One historian has described seventeenth-century England as a union of partially independent counties,

---

[49] 'Devine Providence', Beaulieu Palace House, Papers on Scotch Affairs 3, item 1; J. Hayward, *A Treatise of Union of the Two Realmes of England and Scotland* (London, 1604), 3; *Rapta Tatio: The Mirrour of His Majesties Present Government, tending to the Union of his whole Iland of Brittonie* (London, 1604), sig. F2ᵛ; Thornborough, *Discourse*, 9–10;

[50] For a similar comparative assessment see P. Zagorin, *Rebels and Rulers: 1500–1660* (Cambridge, 1981), ii. 131–2.

[51] N. L. Matthews, *William Sheppard: Cromwell's Law Reformer* (Cambridge, 1984), 225.

[52] D. M. Palliser, *The Age of Elizabeth: England under the Later Tudors, 1547–1603* (London, 1983), 5.

while another has referred to it as a confederation of country houses.[53]

The fact remains, however, that England, even with its weak central administration and strong localism, was the most unitary state in seventeenth-century Europe. If the United Kingdom of Great Britain had been modelled, point for point, upon this English state, the union of England and Scotland would have been very close indeed. Not only would there have been one British parliament and privy council but also one system of British law and one British state church. Englishmen and Scots would have gained complete freedom of trade with each other and they would have become joined in one British nation. Some unionists, most notably James VI and I, hoped that England and Scotland could be united in this complete way. When he and others spoke of 'perfect union', this is generally what they meant. Most of the public figures and writers who dealt with the union question, however, and eventually even King James himself, recognized that perfect union would be difficult to achieve. They therefore modified the unitary model of union either by restricting the proposed union to certain areas of public life or by advocating less than perfect union in one or a number of different areas. The union project of 1706, which resulted in the Treaty of 1707, embodied modifications of the former type. The Treaty established a single British parliament and complete freedom of trade between England and Scotland; to that extent the British state that it created was unitary. On the other hand, the Treaty did not fully integrate the two countries administratively; it did not unite the legal systems or the churches of the two countries; and it did not create anything that resembled a united British nation. To that extent the British state was pluralistic.

The main difficulty in constructing a British union modelled on that of the English unitary state, and the main reason why the English model had to be modified in 1707, was that Scotland was qualitatively different from all the other political

---

[53] A. Everitt, *The Community of Kent and The Great Rebellion, 1640–1660* (Leicester, 1973), 13; H. J. Habakuk, 'England' in *The European Nobility in the Eighteenth Century*, ed. A. Goodwin (London, 1953), 4.

units that had been incorporated into, or otherwise united to, England. Unlike the Duchy of Lancaster, the Palatinate of Chester, and the Principality and Marches of Wales, Scotland in the seventeenth century was a free, independent kingdom, a sovereign state in its own right, possessing its own laws, institutions, and church. It was also a national state, its subjects having had since the fourteenth century a consciousness of their identity as a distinct people. In fact, Scotland was, like England, a unitary state. It did not possess all the characteristics of the English state, for its common law was not as distinctly national as that of England and its church was not as clearly subordinated to the state. The Highlands, moreover, had proved to be much more resistant to efforts at territorial consolidation than the outlying parts of England. But Scotland, like England, possessed a single, if weak, central parliament, a system of common law, a national church, and a national economic policy.[54]

The status of Scotland as an independent, sovereign, and unitary state was the fly in the unionist ointment. It was one thing to bring about a 'perfect' union of a duchy or principality with the unitary English state; it was quite another to unite two states in this way unless one had conquered the other. Nothing better illustrates the magnitude of the problem than the results of the search that the participants in the union debate of the early seventeenth century conducted for a precedent of the type of union that had been proposed. Accustomed to seeking precedents for all difficult political decisions, just as lawyers and judges searched for legal precedents for their actions, public figures and writers gathered an enormous amount of evidence regarding successful unions that had taken place either in the distant or more recent past. Analysis of these precedents revealed that a prefect union of two sovereign states had never taken place except by conquest. There was 'nothing more hard to prove', wrote one contemporary, 'than a perfect union'.[55] The only precedent that appeared to have some relevance to the Anglo-Scottish situa-

---

[54] For a very different view, which stresses Scotland's disunity, see R. Mitchison, _Lordship to Patronage: Scotland 1603–1745_ (London, 1983), 2.

[55] SP 14/7/65, p. 1. See also SP 14/7/64; 14/9A/37 I, f. 123ʳ; HMC, _Laing MSS_, ii. 128.

tion was the union of Poland and Lithuania, which had begun as a personal or regal union in 1386 but which through subsequent negotiations had become an incorporating union. Even this precedent, however, had limited applicability since Lithuania, while technically a state, still ranked as only a grand duchy, not a kingdom, and after the union it accepted the precedency of Poland. Whenever kingdoms had been peacefully united, as had happened in a number of instances in the Iberian peninsula, the individual kingdoms had retained their own laws and institutions. In fact, many of the Spanish kingdoms that had been united by the end of the fifteenth century retained their own identity as states. Their union was essentially a dynastic union, reinforced by the creation of a limited number of joint, conciliar institutions.

The status of Scotland as an independent state also made perfect union more unlikely because it ensured that the union would be negotiated by the representatives of both nations. If Scotland had been a subordinate or conquered political unit in the possession of the English crown, the union could have been achieved by the unilateral action of the English parliament, as had been the case with Wales. Such a union could easily have been perfect, as the union with Wales was, since England could have simply absorbed the subordinate unit and destroyed the institutional basis of its autonomy or semi-autonomy. Since Scotland was a sovereign state, however, a union with England had to be bilateral. Except in 1654, when Scotland was in fact a conquered country, the union had to take the form of a treaty, not an English parliamentary annexation.[56] Theoretically, of course, these negotiations between two independent states could have resulted in a perfect union, but that outcome was most unlikely, for the simple reason that Scotland was not likely to agree to the destruction of its laws and institutions and its complete absorption into an enlarged English state. Throughout the seventeenth century Scotland insisted upon the maintenance of its fundamental laws and institutions in any future union with England. As it turned out, Scotland did eventually

[56] Even then, many Englishmen denied that the union was based on conquest. See, e.g., *The Diary of Thomas Burton, Esq., Member in the Parliaments of Oliver and Richard Cromwell from 1656 to 1659*, ed. J. T. Rutt (London, 1828), iv. 183.

consent to the destruction of many of these fundamental laws, but it did so with two reservations. First, it insisted that those institutions which replaced its own would at least technically be new Anglo-Scottish or British institutions, not simply old English ones. Second, and more important, Scotland insisted that it retain its own system of private, national law and its own church. The union therefore was not perfect, as James had originally hoped it would be, and the new British state was no mere replica or extension of the seventeenth-century English state.

Although the status of Scotland as an independent state seemed indisputable in the seventeenth century, a number of English writers went to great lengths to deny it. Since in the past a number of Scottish kings had paid homage to the king of England, these men argued that England still retained the 'propriety' of the kingdom of Scotland.[57] The argument was predicated upon the false assumption that the lord–vassal relationship that had existed between the kings of England and Scotland in the eleventh, twelfth, and thirteenth centuries had persisted into the seventeenth, despite the failure of any Scottish king to renew the fee in over three hundred years and despite the fact that the English king had not in any way continued to possess Scotland during the same three-hundred year period. The argument had little validity, and a long succession of Scottish writers from Sir Thomas Craig in the early seventeenth century to James Anderson in the late seventeenth effectively refuted it.[58] Nevertheless, the idea of the king of England exercising feudal suzerainty over Scotland did not die easily, and its appeal had considerable relevance to the union question. It shows that although Scotland had all

[57] Bodl., Clarendon MS 133. See also Bodl., Tanner MS 211, ff. 258–290; 'Union by Concurrency of the Homager State with the Superiour', SP 14/7/80X; *Antiquity of England's Superiority; A Convincing Reply to the Lord Beilhaven's Speech in relation to the Pretended Independency of the Scottish Nation* (London, 1706); J. Drake, *Historia Anglo-Scotica: or an Impartial History of all that happen'd between the Kings and Kingdoms of England and Scotland from the Beginning of the Reign of William the Conqueror to the Reign of Queen Elizabeth* (London, 1703).

[58] Sir T. Craig, *Scotland's Soveraignty Asserted* (London, 1695); J. Anderson, *An Historical Essay Showing that the Crown of the Kingdom of Scotland is Imperial and Independent* (Edinburgh, 1705); 'That the Crowne of Scotland was not subject to England', Add. MS 32,094, ff. 247–258. See generally W. Ferguson, 'Imperial Crowns: A Neglected Facet of the Background to the Treaty of Union of 1707', *SHR* 53 (1974), 22–44.

the outward signs of being an independent kingdom that could claim equality of status with England in the various negotiations for a further union, many Englishmen did not wish to consider Scotland in this way. They wished to ascribe to Scotland the same status as Wales in the early sixteenth century and thereby secure a union that was for all practical purposes an annexation of Scotland to England.

Even when the advocates of perfect union did not subscribe to the theory of Scotland's feudal subjection to England, they very often did assign to Scotland a status inferior to that of England and advocated a union in which Scotland, as the inferior partner, would merely be absorbed into a greater England. These men recognized, quite realistically, that if Scotland entered the union as an equal partner with England, the chances of perfect union would be slight indeed. This attitude is most readily apparent in the speeches of that supreme advocate of perfect union, James VI and I. Although James always appeared to be equitable in dealing with Scotland as a partner in the union, and although his critics believed that he wished to give Scotland a predominant role in a united Britain, he always assumed that Scotland would have a minor status within the projected unitary British state and would in effect become nothing more than an outlying English province. In addressing his English parliament in 1607 he told a wary assembly that the projected union would be 'as if you had got it by conquest, but such a conquest as may be cemented by love, the only sure bond of subjection or friendship: that as there is over both but *unus Rex*, so there may be in both but *unus Grex* and *una Lex*'.[59] Continuing in the same vein, he asked rhetorically, 'Must they not be subjected to the laws of England and so with time become but as Cumberland and Northumberland and those other remote and northern shires?' Then, using his favourite marital analogy to describe the union, he insisted that 'You are to be the husband, they the wife; you conquerors, they as conquered, though not by the sword, but by the sweet and sure bond.'[60]

Now, it is certainly true that James would never have spoken these words to his Scottish parliament. Tactless he

[59] *Political Works of James I*, 292.
[60] Ibid. 294.

may have been, but he would never have compared the
kingdom of the Scots to Northumberland, used the term
'conquest' to describe the theoretical foundation of the union,
or proposed the subjection of the Scots to English law if he had
been in Edinburgh. Even in Westminster, speaking to an
exclusively English audience, his use of such terms to describe
the union is surprising. Perhaps James was deliberately
belittling the future position of Scotland within the union in
order to assuage English fears that 'this union will be the crisis
to the overthrow of England and setting up of Scotland'.[61]
Nevertheless, there is no reason to doubt that James was
accurately describing the British state that he hoped would
some day be established. Having decided from the very
moment of his accession, and probably long before, that
Westminster would be the centre from which he would rule
both kingdoms, and being fully aware of the superiority of
England over Scotland with respect to territory, population,
wealth, and military strength, he had no trouble, despite his
Scottish birth and kingship, in adopting a completely Anglo-
centric view of his united kingdoms. He envisaged Scotland
becoming a subordinate, although none the less an important
and integral part of the new state, perhaps like Cumberland,
although probably more like Wales.[62] What James was doing
in this speech of 1607 was not simply appeasing the anger of
his English parliament but describing accurately the position
he planned for Scotland within a unitary British state. There
was just as much reason to take his words at their face value in
this instance as when he prophesied that Scotland, like the
other northern shires, 'will be seldom seen and saluted by
their king, and that as it were but in a posting or hunt-
ing journey'.[63] For the record, it should be noted that
James visited his native kingdom only once after 1603, in
1617.

Somewhat ironically, James received an endorsement of his
view of Scotland's position within a future British state from
his main opponent on the question of union, Sir Edwin
Sandys. The conflict between James and councillors like Sir

[61] *Political Works of James I.*
[62] James wanted Scotland to 'be as Wales was'. Ibid. 212.
[63] Ibid. 294.

Francis Bacon on the one hand and Sandys and his many supporters on the other did not stem from differences regarding long-term goals but from a disagreement over the way in which those goals should be achieved. Aware of the strength of Scottish opposition to an incorporating union with England, James decided in 1604 to achieve union gradually. His first step was to give his support to the articles that English and Scottish commissioners agreed upon in the autumn of 1604. These articles, as mentioned above, would have resulted in the mutual naturalization of Scots and Englishmen and the establishment of limited free trade between them, but they would not have created a single British state. Sandys and his allies objected strenuously to this 'imperfect' union, mainly on the grounds that it would have been unfair to England. Scotland would have acquired the privileges of Englishmen without assuming the corresponding obligations and without becoming subject to the laws made by the English parliament. In the course of the debates on the Instrument of Union Sandys proposed a perfect union, in which Scotland would send representatives to the England parliament and become subject to English law, as an alternative to the imperfect union of the commissioners. The proposal had little chance of acceptance since the Scots would never have consented to the creation of an incorporating union in 1607. In proposing it Sandys may have been simply trying to defeat the imperfect union under discussion.[64] But even if his proposal was nothing more than a tactical manœuver, which is by no means certain, there is no reason to believe that Sandys could not have lived with the type of union he was proposing. At the very least we can say that if there was to have been a union, a perfect union such as proposed in 1607 was the only kind of union that Sandys and his supporters would have been willing to accept.[65]

---

[64] See W. Notestein, *The House of Commons, 1604–1610* (New Haven, Conn., 1971), 233–4, 249, for a discussion of Sandys's motives. See also *L&L* iii. 334–5.

[65] R. E. Ham, *The County and the Kingdom: Sir Herbert Croft and the Elizabethan State* (Washington, DC, 1977), 218. Sir J. Holles, who was less optimistic than Croft regarding the possibility of immediate perfect union, nevertheless agreed that perfect union was the only acceptable type of union, since it provided for the Scots to be 'reduced under our law, without which we have no hold of them neither in love nor safety'. HMC, *Portland MSS*, ix. 110–11, 122–3.

Sandys was, therefore, completely committed to the preservation of the English unitary state. If Scotland were to be united with England, it would have to be as part and parcel of that unitary state. He was not willing, as James and the English commissioners were in 1604, to accept a union on any other terms, even if that imperfect union was intended to be temporary and to provide the foundation for a unitary British state, dominated by England, in the future. From the very beginning Scotland was to be, just as James predicted it would eventually become, a remote region in an enlarged English state.

The position of Sandys and his allies in 1607 was endorsed by many subsequent English parliaments and designated commissioners. Many of these individuals expressed much greater enthusiasm than Sandys for the achievement of union, and many of them were willing to make more concessions than he was to Scottish independence within the union. But throughout the seventeenth and early eighteenth centuries Englishmen demanded, as Sandys had, that if there was to be a union it was to be an incorporating one; that any new British institutions would be essentially English in character; and that Scotland would not be an equal but in many respects a subordinate partner in the union. This was the position that England adopted in the 1650s, when the union was achieved as the result of *de facto* conquest, and again between 1668 and 1707, when the English insisted over and over again on a union of parliaments. The English demand for a 'perfect' or incorporating union explains not only why the English parliament rejected the Instrument of Union in 1607 but also why the English commissioners would not agree to a strictly religious and economic union in 1641 and an exclusively economic one in 1668. Time and time again between 1603 and 1707 the English pursuit of a union such as proposed by Sandys ran up against the traditional Scottish opposition to annexation by, or incorporation into, England. This conflict affected every aspect of the complicated union question of 1603–1707. It was most apparent, however, in the discussion of the strictly political aspects of the question, to which we now turn.

2

Political Union

THE union question touched on every important aspect of English and Scottish life, but it was first and foremost a political and constitutional matter. The most widely discussed and controversial issue that arose as a result of the regal union of 1603 was whether or not and on what terms the bodies politic of the two kingdoms would be joined. This political question dealt with the very structure of the state, the institutions which defined the political community and which possessed the legal authority to compel the obedience of those who lived within the territorial boundaries of the king's dominions. The settlement of this political issue was basic to the broader union question, since the determination of the political structure of Britain could not fail to have a strong bearing on the legal, religious, economic, and social relationships that would be established between the two kingdoms.

In the most general terms the political question was whether there was to be an incorporating union, a federal union, or no further political union at all. To use the models of other united states that contemporaries cited so frequently, the question was whether Britain was to have a political structure resembling that of England, the United Netherlands, or Spain. More specifically, the union question dealt with three sets of problems. The first of these concerned parliament. Was there to be a union of the English and Scottish parliaments, and if so, on what terms? If there was to be a common British parliament, should there also be separate English and Scottish assemblies? How many Scottish representatives would be sent to the new British parliament? Would the distribution of seats be based on population, the number of existing constituencies, or the distribution of wealth in the form of taxable land? The second set of problems concerned the monarchy. Would there be two crowns or one? What would the position of the monarch be in the new British state and how broad would his prerogative be? Finally, how much joint administration would be established? Would certain Scottish offices disappear and be assimilated to those of England? And what restrictions, if any, would be imposed on the holding of office by Scots or Englishmen in the others' country?

Like most constitutional questions, these were not easily capable of solution. Their difficulty derived not only from the determination of each country to achieve the maximum possible strength and influence within the union but also from the unwillingness of both Englishmen and Scots to make any changes in their constitutions. As Sir Thomas Edmondes wrote in September 1604, just before the union commissioners met, 'Most men are of opinion that there be so great difficulty to change the state of the present constitutions, as it is thought that little can be done to satisfy that which is proposed.'[1] The sentiment to which Edmondes referred proved to be one of the main impediments to union during the entire seventeenth century. In some cases it led to opposition to all forms of political union. More commonly, however, it led Scots to demand nothing more than a federal union and Englishmen nothing less than an incorporating one.

I

The fear that union would bring about undesirable constitutional change was much deeper and more widespread in Scotland than in England. Scotland, as the smaller and less populous nation, was much more likely than England to have its political institutions absorbed into those of the other country. For this reason many Scots believed as early as 1604 that the union might destroy their native constitution.[2] In order to prevent this from happening, the members of the Scottish parliament instructed the union commissioners whom they appointed in 1604 not to tolerate any derogation of

[1] Sir R. Winwood, *Memorials of Affairs of State in the Reigns of Q. Elizabeth and K. James I* (London, 1725), ii. 32.
[2] SP 14/8/9–10, 14/22/17; *RPCS* vii. 98; HMC, *Salisbury MSS*, xvi. 98; *Leters and State Papers during the Reign of James VI*, ed. J. Maidment (Abbotsford Club, 1838), 60.

the 'fundamental laws, ancient privileges, offices, rights, dignities and liberties' of their kingdom.[3] It is uncertain exactly which laws the Scottish parliament considered to be fundamental. The term almost certainly applied to the *jus regium* or royal prerogative, which was the meaning James VI gave to it, and it also applied just as readily to the auld laws and to acts of parliament such as that of 1575 which declared it treasonous to alter the structure of parliament.[4] In general usage, the term fundamental law became synonymous with the constitution, or with what Daniel Defoe late referred to as the 'fundamentals of government'.[5] In 1706 Captain Alexander Bruce declared in an anti-union treatise that there were only two fundamental laws, both of which were in danger of being overturned. The first was the kingdom itself and its ancient name of Scotland, under which Bruce subsumed the right to have its own parliament and the right of Scottish peers to sit in that parliament. The second was the 'royal succession to the Imperial Crown of this Kingdom'.[6] According to Bruce, therefore, the fundamental laws were the basic principles of the Scottish constitution, and in fearing their destruction Bruce was merely echoing the prediction of Lord Belhaven that the union would maintain the English constitution intact while subjecting the Scottish constitution to 'regulations or annihilations'.[7]

Scots who protested that an incorporating union would violate fundamental law were much more concerned with the danger that Scotland would lose its sovereignty than with the possibility that it would experience any constitutional change whatsoever. As early eighteenth-century unionists were quick to point out, many of the same men who appealed to fundamental law in arguing against the union had

[3] *APS*, iv. 264.

[4] SP 14/7/53; J. W. Gough, *Fundamental Law in English History* (Oxford, 1955), 48–65; [G. Ridpath], *A Discourse upon the Union of Scotland and England* (London 1702), 46.

[5] D. Defoe, *The History of the Union between England and Scotland* (London, 1786), 361. For a similar usage earlier in the century see *Information from the Estaits of the Kingdome of Scotland to the Kingdome of England* (n.p., 1640), 12.

[6] [A. Bruce], *A Discourse of a Cavalier Gentleman on the Divine and Humane Laws with Respect to the Succession* (n.p., 1706), 3–5.

[7] Daiches, *Scotland and the Union*, 148. See also [A. Fletcher], *State of the Controversy betwixt United and Separate Parliaments* (n.p., 1706), 13.

enthusiastically supported alterations in the fundamental law
of the kingdom in the first Scottish parliament of Charles II
and again in 1689 when parliament expelled the bishops.[8] The
danger was not so much tampering with the law but 'the total
subversion of all the glorious structure our predecessors have
built'.[9] It was also apparent that many Scots were willing to
tolerate some constitutional change in connection with the
union so long as Scottish sovereignty could be preserved.
Those who supported a federal union, for example, were
willing to cede some functions to a joint parliament or council,
a move that would certainly have altered the fundamental law
of Scotland but would still have left the country with its traditional system of government and would have prevented it, at
least formally, from being absorbed into an enlarged English
state.

The Scottish fear of a loss of sovereignty was often expressed
as a fear of conquest. The memory of earlier English attempts
at conquering Scotland and the persistence of English claims
of feudal suzerainty over Scotland made Scots wary, even
before the Union of the Crowns, that union would in effect
place Scotland in the same position as if she had been
conquered.[10] After 1603 James did little to assuage these fears.
Not only did he use the language of conquest to describe the
anticipated nuptials between the two kingdoms but he also
expressed the hope that Scotland would 'be as Wales was', a
statement that could easily remind Scots of the original foundation of Wales's subjection to England.[11] And although James
denied that he ever intended to rule Scotland in the way that
the king of Spain ruled Naples, his Scottish council none the
less expressed concern that His Majesty's ancient and native
kingdom might be turned into a 'conquered and slavish
province to be governed by a Viceroy or Deputy'.[12]

Scotsmen's fears that the union would result in the loss of
their country's sovereignty, the destruction of their constitu-

---

[8] *A Discourse Concerning the Union* [Edinburgh, 1706], 5.

[9] [J. Clerk], *A Letter to a Friend, Giving an Account how the Treaty of Union has been received here* (Edinburgh, 1706), 6.

[10] Sir J. Harington, *A Tract on the Succession to the Crown (A.D. 1602)* ed. C. R. Markham (London, 1880), 19

[11] *Political Works of James I*, 212.

[12] *RPCS*, vii. 536.

tion and treatment by England as if they had been conquered were all borne out in the 1650s, when Cromwell brought about an incorporating union of the two countries. Not only did the Scots recognize that their constitution had been completely changed by the union, but they also could not ignore the fact that the union was based upon conquest and that it was maintained by English military garrisons.[13] The Cromwellian Union did not last, but anti-unionists in the late seventeenth century could easily point to it as an example of the English effort to 'have made us slaves by garrisons'.[14] Even in 1707, when a union was finally established on the basis of a formal treaty, anti-unionists argued that England had in effect conquered Scotland. A Scottish presbyterian tract written in 1707 asserted that the Treaty, which constituted a 'visible and plain subversion of the fundamental ancient constitutions, laws, and liberties of this kingdom', made Scotland, which had never been fully conquered in 2,000 years, 'debased and enslaved'.[15]

If the Scottish fear of constitutional change, which amounted to a fear of losing sovereignty and being conquered, is completely understandable in light of Scotland's position relative to England, the existence of a corresponding fear within the English political nation is somewhat more perplexing. Just like the Scots, many Englishmen claimed, at least in the early seventeenth century, that the union would destroy their fundamental laws, deny England its sovereignty, and result in its conquest. In 1602 Sir John Harington observed that whereas in the time of Protector Somerset the Scots feared a conquest in the guise of a dynastic marriage 'some English fear the like now, foolish fears of men that commonly draw on by fearing that which they most fear'.[16] As Harington suggested, English fears of 'a conquest' were greatly exaggerated, but they still merit close attention, both because they contribute to our understanding of English anti-union sentiment and because they illuminate the general political mood that prevailed in early seventeenth-century England.

[13] This was recognized at Dalkeith in 1651. See *Cromwellian Union*, p. xxix.
[14] Lee, *The Cabal*, 57; Daiches, *Scotland and the Union*, 113.
[15] *A Protestation and Testimony against the Incorporating Union with England* [n.p.,1707], 5.
[16] Harington, *A Tract on the Succession*, 19.

English fears of constitutional alteration or subversion as the result of the union were more complex and varied than those of the Scots. In fact, the prospect of union elicited a number of different apprehensive responses. The first bears the closest resemblance to Scottish fears of *de facto* conquest and absorption into England. As incongruous as it may appear, a number of Englishmen did apparently fear that England might be absorbed into Scotland. Alarmed by the report that a group of Scottish nobles were considering England to be 'accessory to Scotland' because it had fallen to it by inheritance, they worried that the Scots would 'in the next age rule, subdue and in time supplant the English nation'.[17] They feared that Scotland would become the 'predominant part' of a united Britain and that the union would result in the 'overthrow of England and the setting up of Scotland'.[18] How Scotland could ever have achieved this supremacy was not made clear. The most likely possibility is that the king would have appointed Scots to political office in England and that they would have ruled England in Scotland's interest.[19] Whether any Scots actually had such ambitions is uncertain, but the king's appointment of four Scots to the English privy council in 1603 certainly made some Englishmen apprehensive.

While the possibility that Scotland would absorb England or at least make England accessory to it generated a fear of *de facto* conquest and a loss of sovereignty, the method by which Scottish supremacy might have been achieved—the appointment of Scots to political office in England—raised the fear of another constitutional change: the extension of the royal prerogative. The presence of this fear and its negative effect on the achievement of union became most apparent in 1607, when the English parliament debated the clauses in the

[17] *CSP Ven. 1603-7*, 106, 153; *CJ* i. 359; Harl. MS 1305, f. 26; J. P. Kenyon, *Stuart England* (Harmondsworth, 1978), 52. For reference to the fear that the crown of England would be alienated to the Scottish line see Gonville and Caius College, Cambridge, MS 73/40, f. 191ᵛ.

[18] A. Wilson, *The History of Great Britain, being the Life and Reign of King James the First* (London, 1653), 25; *Political Works of James I*, 294. For the exploitation of similar fears by the English royalists of the 1640s see R. Ashton, *The English Civil War: Conservatism and Revolution, 1603-1649* (London, 1978), 213.

[19] D. Stevenson, *The Scottish Revolution, 1637-1644: The Triumph of the Covenanters* (Newton Abbot, 1973), 20; Notestein, *House of Commons*, 224.

Instrument of Union of 1604 that dealt with naturalization. In order to strengthen the chances of the Instrument's acceptance, James had agreed to the inclusion of a clause in the document that excluded Scots from positions of judicature in England and seats in the English parliament until the union was perfected. In making this concession James had accepted a limitation on his prerogative, which allowed him both to make whatever appointments he wished and to grant denizen status to foreigners. As far as James was concerned the concession was a means by which 'the whole people of England may discern the true difference between a crafty tyrant and a just king'.[20] Nevertheless, in order to preserve in principle the prerogative that he was apparently limiting, James secured the inclusion in the treaty of a saving clause allowing him to make the appointments and denizations if he so wished.[21]

This 'salvo to the King's prerogative', as it was later referred to,[22] became a bone of contention in the English parliamentary session of 1606–7. An extensive commentary on the proposed treaty, apparently taken from speeches of MPs in that year, includes a blistering attack on the king's reservation. There were two main objections. First, the clause seemed 'to quicken the prerogative and add a kind of new strength and vigour to it'. Even the inclusion of the king's promise did not limit the prerogative but strengthened it, for 'if the king say he means not to do such a thing, and we gratefully accept his meaning therein, it purposeth this much (inclusive) that he may do it if he would, though he means it not'. Second, and more important, if parliament were to reserve the king's prerogative in this point, it would likewise allow him a prerogative 'to call Scotland to the English Parliament'.[23] Here in concrete form was an illustration of the fear expressed in the Form of Apology and Satisfaction of 1604, that 'the prerogatives of princes may easily and do daily grow'.[24] The

[20] HMC, *Third Report*, App., 11–12.

[21] James insisted that his promise 'be neither restricted till the full accomplishment of the union or to any certain time'. HMC, *Salisbury MSS*, xvi. 363–4.

[22] 'A Discourse upon the Union of England and Scotland', in *Miscellanea Aulica*, ed. T. Brown (London, 1702), 193.

[23] Harl. MS 1314, p. 77. Thomas Hedley made this last point in the conference on naturalization, 19 Feb. 1607. SP 14/26/54

[24] *Constitutional Documents of the Reign of James I*, ed. J. R. Tanner (Cambridge,

union in other words, could not be dissociated from the more
general concern that many MPs had regarding the king's
powers in the first decade of the seventeenth century.

The fear of an enlarged royal prerogative also contributed to another constitutional fear concerning the union that
Englishmen harboured in 1604, the fear that the union would
create a new kingdom and thereby destroy the English
constitution. This fear first arose in the parliamentary debate
of 1604 over the king's proposal to change the names of his
kingdoms to Great Britain. Against this proposal the
opponents of union raised a long series of objections, with
which many of the pro-union pamphleteers took issue in 1604
and 1605.[25] Many of these objections were contrived and
exaggerated, but at the very least one had to be taken
seriously. This was the objection that 'the change of style will
be, as it were, the erecting of a new kingdom, and so it shall be
as it were, a kingdom conquered, and then may the king add
laws and alter laws at his pleasure'.[26] On this point the anti-
unionists appeared to have the law on their side, for in April
1604 James had asked the judges whether the use of the name
Britain could be warranted by act of parliament without the
direct abrogation of all the laws of England. They had replied
in the negative, claiming that the kingdoms could not be so
united until their laws were united.[27] They said nothing that
would have prevented James from assuming the new style by
his power of proclamation, which he did without legal
objection later that year, but they did present a serious legal
obstacle to making the change by statute.[28]

In delivering this opinion the judges did not, as James had
feared, base their decision on 'the curious wresting of the
common law of England against all reason'.[29] As far as the
judges were concerned, an act changing the names of the two
kingdoms would to some extent actually have united them. As
they defined it, the question was 'whether there could be made

1961), 222. For a foreign observation that the union debate might lead to a reduction
of royal authority and an increase in parliamentary power see *CSP Ven. 1603–7*, 485.
    [25] *Jacobean Union*, pp. xx–xxxv.   [26] Harl. MS 292, f. 133.
    [27] Sir E. Coke, *The Fourth Part of the Institutes of the Laws of England* (London,
1644), 347.   [28] *L&L*, iii. 225–6.
    [29] SP 14/7/38. James also referred to it as a 'foolish quirk of the judges'. *Political
Works of James I*, 329.

a union of the kingdoms by raising a new Kingdom of Great Britain', or as Justice Warburton wrote, 'whether Scotland may be united to England and made parcel of England'.[30] Phrased in those terms, their negative response was completely understandable, since an actual union did require that the laws (or at least the public laws) of the two kingdoms first be united. The concern therefore was not with the mere technicality of a change in the name of the kingdom, but with the reality of the union which the change in the name both symbolized and to some extent achieved.[31] Sir Edwin Sandys, who spoke most strongly against the king's proposal in 1604, made it perfectly clear what the issue was and at the same time revealed the depth of anti-unionist fears when he said, 'The King cannot preserve the fundamental laws by uniting no more than a goldsmith two crowns. The bare alteration of the name taketh them not away, but a union doth. . . . We shall alter all laws, customs, privileges by uniting.'[32]

In dealing with all this anti-unionist sentiment in the first English parliament of James I, one must consider the possibility that all these references to the destruction of fundamental law, the establishment of a conquered kingdom, and the threat of arbitrary kingship were nothing but rhetorical devices, intended to stir up opposition to union that was based on much more vulgar national prejudices. If we did not have other evidence of a profound constitutional fear in early seventeenth-century England, we might be tempted to conclude that that was in fact the case. But the parliamentary record of the early seventeenth century gives abundant evidence of a pervasive fear of constitutional change and subversion. The Form of Apology and Satisfaction was only one of many examples of this mentality, the most dramatic being Edward Alford's statement in the parliament of 1628 that if parliament allowed its privileges to be encroached on any further, then 'farewell Parliament and farewell England'.[33]

[30] SP 14/7/38; Coke, *Fourth Part of the Institutes*, 347; Hampshire Record Office, Jervoise MS 44M69, no. 79.

[31] For a learned opinion that the union could not create a new kingdom see W. Clerk, 'Ancillans Synopsis', Trinity College Dublin, MS 635, p. 20.

[32] *CJ* i. 187.

[33] Quoted in D. Hirst, *The Representative of the People?: Voters and Voting in England under the Early Stuarts* (Cambridge, 1975), 66.

And why should not members of parliament have been apprehensive? James I never really acted like an absolute monarch, but he did often speak and write like one, both before and after he arrived in England. It was perfectly understandable that English MPs should have expressed concern about the growth of the prerogative. It was also understandable that they would have used the language of conquest to describe the situation they wished to avoid. Those who had read Fortescue (and many MPs had) knew that the 'regal' or absolutistic government of France, unlike that of England, had derived from conquest.[34] They also knew from Fortescue that when a king—any king—did not rule according to the law, he could be viewed as a conqueror, in the manner of the biblical Nimrod, and that his authority as conqueror was legitimate.[35] It followed, therefore, that if the creation of a new kingdom abrogated the old laws of England and Scotland, the laws by which James ruled both kingdoms, then he would *ipso facto* rule as a conqueror. The fear of conquest, in other words, was legitimate.

The prevailing English fear that the union would destroy the fundamental law or constitution of England had both short-term and long-term effects. In the short run it contributed to the demise of James's union project, leading directly to the rejection of the proposed statutory change of the name of the kingdom in 1604 and the rejection of most of the articles of union submitted to parliament in 1606. In the long run it encouraged Englishmen to consider an incorporating union as

[34] Sir J. Fortescue, *De Laudibus Legum Anglie*, ed. S. B. Chrimes (Cambridge, 1942), 28; id., *The Governance of England*, ed. C. Plummer (Oxford 1885), 109–16; R. Eccleshall, *Order and Reason in Politics* (Oxford, 1978), 107. According to this point of view, the Norman Conquest was not really a conquest at all, since it had not changed the laws or the frame of government. Q. Skinner, 'History and Ideology in the English Revolution', *Historical Journal*, 8 (1965) 152–3. For the argument that English monarchs 'have it in their power to act in a despotical manner over a conquered people such as Ireland and the Western Plantation' see *Remarks upon a late Dangerous Pamphlet, intitled The Reducing of Scotland by Arms* (London, 1705), 4.

[35] J. G. A. Pocock, *The Ancient Constitution and the Feudal Law* (2nd edn., Cambridge, 1987), pp. 283–5. I am grateful to Professor Pocock for letting me see the new sections of this edition before publication. On the legitimacy of power by conquest see D. Sutherland, 'Conquest and Law', *Studia Gratiana*, 15 (1972), 25. In 1692 an Englishman wrote that 'the power that Conquers will give Laws and Religion to the Conquer'd'. *A Letter to a Friend concerning a French Invasion* (London, 1692).

the only acceptable type of Anglo-Scottish union. The only way to unite Scotland with England without changing the English constitution was to incorporate Scotland into what would remain essentially an English parliament and to make Scotland subject to English statutory law. Even Sandys, the most adamant opponent of James's union project in 1604 and 1606–7, claimed that this type of union would be acceptable and actually proposed such an arrangement, which he referred to as 'perfect union', in 1607. In the 1650s the Cromwellian government established precisely this type of union without raising any of the old constitutional fears, and after 1668 incorporation became virtually a non-negotiable English demand in all union negotiations.[36] It is worthwhile noting that when an incorporating union was finally agreed upon in 1707, the names of the two kingdoms were changed to Great Britain by acts of parliament without too much debate.[37] This would tend to support the view that in the early seventeenth century the source of English constitutional fear was not so much the change of name itself but the situation that would have resulted if the name were to have been changed without bringing about an incorporating union.

Scottish and English fears of constitutional change underlay the entire union debate. They explain a great deal of the opposition to union that arose in both countries and they also explain why a significant segment of Scottish opinion would consider only a federal union while most English thought favoured an incorporating one.[38] But in order to appreciate the complexity and difficulty of the constitutional question, and to discover why the settlement reached in 1707 took the form that it did, we must turn to the specific questions that were debated in the midst of such great apprehension.

[36] As late as 1707, however, one Englishman, Lord Haversham, objected to an incorporating union (but not a federal union) because he thought it would destroy 'the good old English Constitution'. J. Thompson, Baron Haversham, *The Lord Haversham's Speech in the House of Peers on Saturday, February 15, 1706–7* (London, 1707), 2.

[37] Pryde, *Treaty of Union*, 83, Art. I. The acts of parliament were the English and Scottish Acts of Union which gave the Treaty its judicial basis.

[38] For unionist attempts to dispel these fears in the early 18th century see W. Seton, *A Speech in Parliament the second day of November, 1706* (Edinburgh, 1706), in Defoe, *History of the Union*, 361.

II

The first of these questions dealt with the union of parliaments. Somewhat surprisingly this central issue did not assume paramount importance until the middle of the seventeenth century. It is true that James VI and I definitely had a union of parliaments in mind when he first proposed a perfect union of the two countries. One of the topics that he submitted to the union commissioners for discussion in September 1604 was 'the reducing of both parliaments into one', while two months later he still spoke of parliamentary union as at least a distant goal.[39] After that date, however, James apparently abandoned his plans for uniting the parliaments, mentioning only once in 1607 that he hoped to find a mean between the short Scottish and long English parliamentary sessions.[40]

Historians have often attributed James's volte-face on the question of parliamentary union to the difficulties he encountered in dealing with his English parliament. Better to keep the more manageable Scottish parliament separate, so the argument goes, than to compound his difficulties in England.[41] The only problem with this interpretation is that James had already run into trouble with his English parliament before his proposal of September 1604. It is much more likely that James abandoned the prospect of parliamentary union when he realized that the commissioners would not, and in fact the Scottish commissioners could not, undertake such fundamental constitutional change. James, of course, did not consider himself tied to the recommendations of the commissioners,[42] but he soon realized that the imperfect, non-parliamentary union they proposed was the most he could hope to achieve, at least in his lifetime, and therefore he scaled down his objectives accordingly.

James might also have been influenced by some of the union tracts written in 1604. One in particular, a scholarly treatise by Sir Henry Savile, the warden of Merton College, Oxford, might have carried considerable weight with James, since

---

[39] SP 14/9/35A; Huntington Library, Ellesmere MS 1225; HMC, *Salisbury MSS*, xvi. 363.
[40] *Political Works of James I*, 300.
[41] H. R. Trevor-Roper, *Religion, the Reformation and Social Change* (London, 1967), 451 n.   [42] HMC, *Salisbury MSS*, xvi. 395.

the king himself, or one of his ministers, had specifically commissioned Savile to write it while the king was at Windsor in September 1604.[43] Like most other authors of early seventeenth-century union tracts, Savile spoke the language of perfect union, but on the crucial question of parliamentary union he made an emphatically negative recommendation, citing as precedents the reluctance of other states to enter into similar arrangements.[44] Savile was not alone in his opinion. Sir Thomas Craig, an enthusiastic supporter of further union who served as one of the Scottish commissioners in 1604, also recommended that the parliaments be kept separate. For Craig the entire question of parliamentary union was not of great importance, and it was not necessary to achieve it in order to bring about an 'inviolable' or 'incorporating' union.[45]

As it was, very few unionists of the early seventeenth century thought in terms of parliamentary union. Preoccupied as they were with the regal character of the Union of the Crowns, they reflected mainly on the already accomplished union in the king's person and the strength which that union derived from geography, language, and religion. Some of these authors, to be sure, spoke of a union in the body as well as in the head and looked forward to a union of laws and of love. Yet oddly enough, very few of them regarded parliamentary union as part of their schemes, even though the English parliament was usually regarded as the most perfect symbol of the English body politic and both parliaments would have played central roles in the achievement of legal union. John Hayward actually devoted an entire chapter to the union of laws without once mentioning either the English or the Scottish parliament.[46] It was left to the anti-unionists to point out the absurdity, if not the danger, of a union of laws without a union of parliaments.[47]

[43] Bodl. MS e Museo 55, f. 93; *Jacobean Union*, pp. lxxv–lxxvii.

[44] Sir H. Savile, 'Historicall Collections', in *Jacobean Union* 233–4. Only one of the many unions he discussed, the union of Poland and Lithuania, had led to a union of diets or parliaments.

[45] Sir T. Craig, *De Unione Regnorum Britanniae Tractatus*, ed. C. S. Terry (SHS, Edinburgh 1909), 282–3.

[46] Hayward, *A Treatise of Union*, 8–11.

[47] Smout, 'Road to Union', 177, *Bowyer Diary*, 265. The same danger appeared with respect to commercial union. See Keith, *Commercial Relations of England and Scotland*, 14, 46, Notestein, *House of Commons*, 235.

The question then arises why James, who had once supported a union of parliaments, objected so strenuously to the perfect union that Sandys and his cohorts proposed in 1607, for according to that plan Scotland would have been fully incorporated into the English body politic. To some extent James took the position he did because he did not wish to restrict the prerogative. One of the main reasons why Sandys made his proposal was to resolve the difficult question of the *post-nati* and the *ante-nati* by bringing all Scots under the direct control of English law.[48] This would have denied James his right 'to denizate, enable and prefer to such offices, honours, dignities, and benefices whatsoever' in the case of the *ante-nati* Scots.[49] The king had, of course, agreed to accept limitations on this power, but he wished to retain that option, since 'it is a special point of the king's own prerogative to make aliens citizens'.[50] Another reason for James's opposition to Sandys's proposal was that by 1607 he had come to the conclusion, probably through the influence of Bacon, that a union of laws would require extensive preparation and should not be implemented immediately.[51] The plan of Sandys, in other words, was not only constitutionally restrictive but also precipitate.

Sandys's proposal, of course, got nowhere, and between 1608 and 1651 there was no further consideration of the possibility of parliamentary union. Even when the two parliaments entered into negotiations for some sort of union between the two nations in the 1640s, they concentrated on plans for religious unity and joint executive action, not a union of the two legislatures. It was not until the Cromwellian conquest of Scotland and the subsequent negotiations at Dalkeith and London that full attention was finally paid to the possibility of parliamentary union. The need to control Scotland, which had taken up arms against the Independents in the second Civil War, made a Welsh-style annexation of the country imperative, at least in the thinking of Cromwell and his colleagues. Since England was by that time a republic, in which parliament clearly constituted the highest authority in

---

[48] *Bowyer Diary*, 257 n.   [49] Defoe, *History of the Union*, 724.
[50] *Political works of James I*, 299.
[51] Sir F. Bacon, 'A Brief Discourse touching the Happy Union of the Kingdoms of England and Scotland', *L&L* iii. 98; *Political Works of James I*, 293.

the realm, an incorporating union could only be achieved by uniting the two parliaments. The Restoration, while destroying the Cromwellian Union, did nothing to diminish the English belief that if a union were to be negotiated, it would have to be a parliamentary one. The failure of the attempt to conclude an economic treaty with Scotland without uniting the two parliaments in 1668 only strengthened this opinion, and after 1670 all formal discussion of the union question involved a consideration of parliamentary union.

As the possibility of parliamentary union became more likely, the determination of the size of the future British assembly and the proportion between English and Scottish members took on much greater urgency. Bacon had posed the question in 1604 when James was still giving serious consideration to a union of parliaments,[52] but a body of thought on the subject did not develop until the 1650s, when the English commissioners limited the Scottish delegation to parliament to thirty members on the basis of relative taxable wealth.[53] The Scottish commissioners protested vehemently against this figure, arguing that the proportion should be based on the relative number of parishes and the relative intrinsic (rather than taxable) value of those parishes. These protests were to no avail, but in the negotiations of 1670 the Scots resumed their bargaining for a larger delegation than the English were willing to concede. The original plan presented to the commissioners in 1670 called for representation on the basis of the Cromwellian precedent. There were to be thirty Scottish MPs in the House of Commons and ten Scottish peers and two bishops in the restored House of Lords. For reasons that are still not completely clear, the Earl of Lauderdale, acting either as a saboteur of the negotiations or as the representative of a wary Scottish commission, demanded that all the members of the Scottish parliament be admitted to the new assembly. Whatever his motives, this demand was finally responsible for bringing the union negotiations of that year to an unsuccessful conclusion.[54]

[52] Sir F. Bacon, 'Certain Articles or Considerations touching the Union of the Kingdoms of England and Scotland', *L&L* iii. 228–9.

[53] *Cromwellian Union*, p. xli.

[54] Ibid., 188; Lee, *The Cabal*, 67; Ferguson, *Scotland's Relations with England*, 156.

The Scottish demand for full representation was advanced again in the union debates of the early eighteenth century. In this enterprise the writings of Sir George Mackenzie of Rosehaugh played a prominent part. Mackenzie, who is best known for his harsh treatment of the Covenanters in 1685 and his authoritative writings on Scottish criminal law, also wrote extensively about the negotiations of 1670, both in his *Memoirs of the Affairs of Scotland* and in *A Discourse Concerning the Three Unions Between Scotland and England*. In the latter work he insisted that James VI had never seen fit to press for a union that would have destroyed the sovereignty of either kingdom, and he quoted the Scottish act of 1604 that had insisted upon the preservation of Scottish fundamental law.[55] In the same vein he wrote an extensive commentary on this act, the main thrust of which was that a union of parliaments contradicted fundamental law, but that if such union were to be negotiated, it must include all the members of the Scottish parliament. This commentary was printed by Robert Wyllie in 1706, long after Mackenzie had died, together with the opinion of Sir John Nisbet, who had strongly opposed the union of parliaments in 1670.[56] In this way Mackenzie's position on Scottish representation became the rallying cry of those Scots who wished to preserve as much as possible of the old Scottish parliament within the proposed new British state. The Scottish commissioners of 1706, however, did not adopt Mackenzie's position, arguing instead for a Scottish delegation of fifty members after the English had proposed thirty-eight. Both figures were significantly higher than the number of MPs to which Scotland would have been entitled on the basis of taxable wealth, for if the returns on the land tax had been used as a standard, Scotland would have received only thirteen members. As it was, the commissioners settled on a figure of forty-five MPs and sixteen peers, who were to be elected by the full body of the Scottish nobility.[57] This was hardly what the commissioners of 1670 or Mackenzie had

---

[55] NLS, Advocates MS 31, 7, 7, ff. 136–7. The *Discourse* was published in 1714.

[56] [Wyllie, *A Letter concerning the Union*. See also [D. Symson], *Sir George M'Kenzie's arguments against an incorporating union particularly considered* (Edinburgh, 1706).

[57] Daiches, *Scotland and the Union*, 131.

demanded, but it was significantly higher than the number of representatives the Scots had received in 1654.

Closely related to Scottish proposals for complete Scottish representation in a united Parliament were those for the maintenance of two separate parliaments which would delegate certain functions to either a united British parliament or council. These schemes are often labelled federal, although technically they were confederative, since the powers of the central parliament would have been delegated from the national assemblies and not, as in a strictly federal arrangement, clearly distinguished from those of the national bodies by a written constitution.[58] These schemes for 'federal' union came exclusively from Scotland in the early eighteenth century, but they actually had their origin in a number of English unionist writings in the early seventeenth century. Indeed, the two English writers of the early seventeenth century besides Bacon who gave serious consideration of a union of parliaments both recommended this type of arrangement.

The first of these authors was John Doddridge, who in 'A Breif Consideracion of the Unyon' discussed the 'discommodities' that might attend the union. Assuming that there could not be a perfect union unless a common parliament was established, Doddridge expressed concern that in such a joint body there would be an inequality of representation. In this way Doddridge anticipated the concerns of Mackenzie and Nisbet more than a half-century later. What is most interesting about Doddridge's proposal, however, is his assumption that in addition to the joint British parliament each country would retain its own separate parliament, just as the cantons of Switzerland had.[59] Although he used the Swiss example, he could just as easily have referred to the confederative structure of the nascent United Provinces of the Netherlands or even that of the Holy Roman Empire, as one early eighteenth-century writer did in his commentary on Doddridge's scheme.[60]

[58] V. Bogdanor, *Devolution* (London, 1979); K. C. Wheare, *Federal Government* (4th edn., London, 1963), 43. See also N. MacCormick, 'Independence and Constitutional Change', in *The Scottish Debate*, ed. N. MacCormick (London, 1970), 55.

[59] J. Doddridge, 'A Breif Consideracion of the Unyon', in *Jacobean Union*, 146.

[60] *Vulpone*, 14.

While Doddridge referred only in passing to the idea of federal union, another anonymous early seventeenth-century Englishman drafted a more complete constitutional proposal for such an arrangement, calling not only for the perpetuation of 'national parliaments' but also for the establishment of a 'general' parliament to handle matters of common concern. All legal matters would be restricted to the national assemblies, unless there was conflict between them, in which case the general parliament would settle the matter. Individuals born in one country who possessed land or dignities in the other would be entitled to sit in the national parliament of the other. This federal scheme corresponded to a similar plan for ecclesiastical government, in which a system of national and general synods would be established.[61]

Neither this 'federal' constitution nor the proposal by Doddridge appears to have won much support in the early seventeenth century, either in England or Scotland. But both of them did provide statesmen in the future with an alternative to the English model on the one hand and the Spanish model on the other. Whether either of these early plans significantly influenced later thought on the subject is uncertain, although Doddridge's plan was certainly known and cited in the early eighteenth century. It is unlikely that either plan provided much guidance in the 1640s, when both countries considered schemes that would facilitate their co-ordination of the war effort and promote religious unity, since none of these plans included provisions for a union of parliaments. The only 'federalism' that was contemplated at that time was the loosest type possible, according to which a special commission, such as the Committee of Both Kingdoms in 1644, would handle common concerns. In the late seventeenth and early eighteenth centuries, however, there was much more talk of a federal solution to the union problem. In the negotiations of 1670, for example, Lauderdale raised not only the possibility of the entire Scottish parliament being combined with the English parliament but of the two sitting separately unless there was an emergency, in which case they would sit jointly.[62]

---

[61] 'The Devine Providence in the Misticall and Real Union of England and Scotland', Beaulieu Palace House, Papers on Scotch Affairs, iii, item 1.

[62] Mackenzie, *Memoirs of the Affairs of Scotland*, 206–8; *Cromwellian Union*, 188.

This proposal represented a curious variation on the theme of federal union. Other plans, more similar to that of Doddridge, appeared at the time of the Glorious Revolution and during the debates of the early eighteenth century. While differing in detail, these plans all advocated the establishment of some sort of joint body to deal with matters of common concern without destroying the separate parliaments of the two nations.[63]

Many of the Scots who argued for a federal union in the early eighteenth century were not, however, thinking either of a 'general' parliament that would supplement the national assemblies or of joint parliamentary sessions. When anti-unionists like James Hodges spoke of a federal union they meant a much stricter form of confederation according to which the legislative efforts of the two separate national parliaments would be co-ordinated by a joint preparatory committee. Each parliament would possess a veto power over such joint measures, a provision that ensured that each country would remain a 'free state'. The real cement of the union would be allegiance to one monarch, but even that was conditional, since Hodges envisioned a number of circumstances in which Scotland would have to withdraw from that allegiance in its own interest.[64] Hodges claimed that his federal union constituted more than a mere con-federation, since the two monarchies would be fully united or 'incorporated'.[65] In fact, however, his proposed union was extraordinarily loose, being little more than what the Scots had proposed in the 1640s. It bore a closer resemblance to the union of the Spanish kingdoms (which Hodges oddly considered to be an incorporating union) than to the federal union of the Netherlands, for in the latter country there was a general political assembly, the States General.[66] Hodges's plan

[63] *Vulpone*, 14; Daiches, *Scotland and the Union*, 133; [Ridpath, *A Discourse upon the Union*, 89–91. See also [F. Grant], *The Patriot Resolved in a Letter to an Addresser* (n.p., 1707), 5.

[64] [J. Hodges], *The Rights and Interests of the Two British Monarchies, Treatise I* (London, 1703), 6.

[65] Ibid. 6–7.

[66] Hodges discusses many foreign unions but confuses the issue by designating the union of the Spanish kingdoms as an incorporating union and the union of Poland and Lithuania as a federal one. Ibid. 4.

for 'federal union' comes closest to that proposed by Thomas
Jefferson seventy years later between the American colonies
and Great Britain, according to which several independent
legislatures would be bound to one king.[67]

Although the idea of federal union achieved great popularity in Scotland in the early eighteenth century, it was given
very short shrift by the union commissioners of 1706. The
English commissioners were, as expected, adamantly opposed
to it and would not consider anything but an incorporating
union.[68] Underlying their position was a deep distrust of
federalism, which during the republican period had been
referred to contemptuously as 'cantonizing'.[69] More specifically, the commissioners had doubts about the main problem
that all federal systems encounter: the duplication of functions
and consequent conflict between regional and national, or
national and general, authorities. It was for this practical
reason that the Duke of Marlborough objected to the federal
constitution of the United Provinces of the Netherlands.
Another reason for opposing federal union was that if
Scotland were to retain its own parliament, it would have the
capacity to undo the union and thereby deny England the
security which was their main motive for entering the union.
According to Hodges's scheme this possibility was explicitly
recognized, but it was implicit in all federal solutions, since
the main purpose of retaining a Scottish parliament was to
preserve Scottish sovereignty or independence within the
union.

The Scottish commissioners of 1706 also rejected
federalism, despite its popularity in Scotland, but their
reasons differed somewhat from those of the English. For the
Scots the problem was that a separate Scottish parliament
could not remain sovereign in any federal arrangement. As

[67] A. M. Lewis, 'Jefferson's *Summary View* as a Chart of Political Union', *William and
Mary Quarterly*, 3rd ser. 5 (1948), 34–51.

[68] In 1706 the Earl of Mar observed that the English 'think all the notions about
federal unions and forms a mere jest and chimera'. *State Papers and Letters Addressed to
William Carstares*, ed. J. McCormick (Edinburgh, 1774), 743–4.

[69] C. Hill, *God's Englishman: Oliver Cromwell and the English Revolution* (London,
1970), 178. For an early 18th-century unionist view of the weakness of the Swiss
system see H. Chamberlen, *The Great Advantages to Both Kingdoms of Scotland and England
by an Union* (n.p., 1702), 8.

they saw it, once a federal system was implemented, sovereignty would have to be located in some institutional body. In the Netherlands it was in the States General; in Britain it would be in the joint assembly or parliament that was established by the union. In other words the hope of Hodges that each nation would remain a free and sovereign state could not be realized. The commissioners 'considered it [federal union] ridiculous and impracticable, for that in all the federal unions there behoved to be a supreme power lodged somewhere, and wherever this was lodged it henceforth became the States General or, in our way of speaking, the Parliament of Great Britain'.[70] In formulating this argument the Scottish commissioners revealed that they were incapable of thinking in terms of divided sovereignty, as many North Americans did later in the century and as proponents of federalism do today.[71] In order to do this it would have been necessary for them to posit co-ordinate local and federal authorities, each of which would possess attributes of sovereignty, the division between them being clearly set down in a written constitution. In Britain in the early eighteenth century this proved to be impossible. Either the separate nations of England and Scotland would be completely sovereign or the joint, British institution established by the union would be. The Scottish commissioners decided that the latter situation would prevail.

III

The second major set of constitutional problems that had to be resolved in connection with the union concerned the crown itself. Paramount among these was whether or not there was to be one imperial crown of Great Britain or two separate crowns of England and Scotland. Some historians would argue that this was really not an issue during the period 1603–1707, since the crowns had already been united in 1603. No one contests the fact that James VI and I united the crowns in his person through joint possession, but this union does not

---

[70] Daiches, *Scotland and the Union*, 129. For a similar argument, see *The Advantages of Scotland by an Incorporate Union with England* (n.p. 1706), 21.

[71] MacCormick, 'Independence and Constitutional Change', 57.

necessarily mean that there was one crown. It is true that the English parliament in the first statute of James's reign declared that the two kingdoms had been united under 'one Imperial Crown'.[72] James frequently reminded his English subjects of this parliamentary recognition, and references to a single imperial crown appear in the writings of many early seventeenth-century unionists.[73] Even Lord Ellesmere, in giving his opinion in *Calvin's Case*, accepted the statement of the English parliament in 1604 as a valid commentary on the constitutional implications of James's accession of the English throne.[74]

Although Englishmen made frequent reference to a single British crown after 1603, they would have had great difficulty explaining the process in constitutional terms. There are three different ways in which one can view the English crown in the early seventeenth century. The first is as a ceremonial crown, the jewelled headpiece worn on state occasions. There was clearly no union of these two crowns in the early seventeenth century, although Bacon recommended that King James consider the means to accomplish that end.[75] Neither he nor his successors did anything along such lines, however, and Hodges could therefore accurately report in 1706 that early seventeenth-century proposals for 'melting down the two crowns and making one out of both' had failed.[76]

A second way of looking at the crown is as the kingship itself, a personal, dynastic inheritance that the king acquired at the time of his accession. The two crowns, conceived of in this fashion, were not united in 1603. James inherited the two crowns at different times and possessed them jointly, knowing that different laws of succession in England and Scotland could separate them again in the future. Until both crowns came under the same laws of succession, as they did in 1707, there could be no single crown. The only way that one could argue that the two crowns were united in this sense was to claim that the king possessed feudal suzerainty over Scotland,

[72] I Jac. I. c. I.
[73] *Stuart Royal Proclamations*, i. 95. J. Thornborough chose to use the expression 'one imperial monarchy', *Ioiefull and Blessed Reuniting*, 7.
[74] *State Trials* ii. 566.
[75] *L&L* iii. 224–5.
[76] Hodges, *Rights and Interests, Treatise 1*, 22

in which case James as king of England merely regained in 1603 what had always been his possession.

The third way of viewing the crown in the early seventeenth century is as a 'corporation sole', a legal fiction denoting the king's body politic.[77] According to contemporary political thinking, this body, which was synonymous with the kingdom itself and could never die, was inseparably united to the king's natural, mortal body. To claim that the two bodies were separable was in fact treasonous, for such a theory would allow a subject to rebel against the natural body of a king by claiming loyalty to his political body. The Despensers had used this very idea to justify their rebellion against Edward II, and parliamentarians later had recourse to the same notion in 1642. The theory of inseparability might have led some Englishmen to think in terms of one united crown. If there was only one king in Britain, and if it was treasonous to draw a distinction between the king's natural body and the crown, then it would follow that there should be only one crown or kingdom. Ellesmere spoke in these terms when he concurred with the judges' decision in *Calvin's Case*.[78] According to Sir Edward Coke, however, the regal union did nothing to unite the crowns or kingdoms. To be sure, there was one king to whom Englishmen and Scots are born into allegiance, but that same king had two different political capacities.[79] This meant, in effect, that there were still two crowns, either of which could be referred to as 'imperial'.[80]

The possibility that Scotland might break the regal union

[77] F. W. Maitland, 'The Crown as Corporation', *Law Quarterly Review*, 17 (1901), 131–46; E. Kantorowitz, *The King's Two Bodies: A Study in Mediaeval Political Theology* (Princeton, NJ, 1957); R. Nevo, *The Dial of Virtue: Poems on Affairs of State in the Seventeenth Century* (Princeton, NJ, 1963), 24, 42. In a very real sense the king was the state, but by using this language the state was personified.

[78] Lord Ellesmere's speech on the *post-nati*, in L. A. Knafla, *Law and Politics in Jacobean England: The Tracts of Lord Chancellor Ellesmere* (Cambridge, 1977), 240, 244–5.

[79] For a discussion of Coke's reasoning in *Calvin's Case* see J. H. Kettner, *The Development of American Citizenship, 1608–1870* (Chapel Hill, NC 1978), 13–28. For the differences between Coke and Ellesmere see D. Hanson, *From Kingdom to Commonwealth: The Development of Civic Consciousness in English Political Thought* (Cambridge, Mass., 1970), 312–15. See also H. Wheeler, 'Calvin's Case (1608) and the McIlwain–Schuyler Debate', *AHR* 61 (1956), 587–97.

[80] See for example [A. Mudie], *Scotiae Indiculum* (London, 1682), Epistle Dedicatory; Sir D. Hume of Crossrigg, *A Diary of the proceedings in the Parliament and Privy Council of Scotland, May, 21, 1700–March 7, 1707* (Edinburgh, 1828), 117.

and establish its own independent monarchy in the early eighteenth century led many Englishmen to seek a more lasting union of the crowns than James VI and I had accomplished.[81] The Treaty of Union achieved this goal, but it did so in a rather indirect way. The first article provided for the union of the two kingdoms into the single kingdom of Great Britain without mentioning the monarchy or the crown. Those words appeared only in the second article, which regulated the succession to the 'Monarchy of the United Kingdom of Great Britain' and the possession of the 'Imperial Crown of Great Britain'.[82] In this way the single crown to which James had frequently referred become a part of the public law of the kingdom.

On one issue relating to the crown, the extent of the royal prerogative, the Treaty of 1707 was silent, although it did never the less have an effect. Modern Scottish writers have expressed regret that the Treaty, which assumed the status of fundamental law, did not place specific limitations on the prerogative and thereby help to define the rights of the subject, or at least the rights of Scottish subjects.[83] However desirable such definition might have been, it is understandable why it did not materialize. Statements of fundamental rights were incorporated into parliamentary statutes of both England and Scotland in the seventeenth century only when it became necessary to remedy specific abuses of royal power. The English Petition of Right (1628), the English Bill of Rights (1689), and the Scottish Claim of Right (1689) were not intended either to establish new rights or to provide a compendium of old rights but simply to give statutory force to specific rights which were being violated and for which legal remedy could not be obtained. The union negotiations did not deal with any such alleged violations of specific rights, and therefore the Treaty did not address such issues. It said nothing about either the prerogative or the individual's rights,

---

[81] In 1702 Blackerby Fairfax formulated a series of proposals for 'a union of Crowns', *A Discourse upon the Uniting Scotland with England* (London, 1702), 62–3.

[82] Pryde, *The Treaty of Union*, 83–4, art. 1–2.

[83] T. B. Smith, 'Scottish Nationalism. Law and Self-Government', in *The Scottish Debate*, 37.

although by leaving both English and Scots private law intact, it maintained the status quo in that regard.

Nevertheless, the extent of the king's power had been a source of considerable debate in connection with earlier union projects. As we have seen, the union debates of those years revealed deep fears that the union would increase the prerogative and diminish liberty. The whole problem was compounded by a great deal of uncertainty regarding the Scottish prerogative, for the king of Scotland could be regarded at the same time as both more powerful and weaker than the king of England. Although he could be considered the most limited prince in Europe at the beginning of the seventeenth century, he possessed a prerogative that the most absolute had not yet attained.[84] The limitations, many of which owed their origin to the strength of the Scottish nobility *vis-à-vis* the king in the Middle Ages, had become part of the Scottish constitution as the nobles had taken steps to cover their rebellious acts with the 'cloak of legality'.[85] Many of these restrictions of royal power continued to operate in the seventeenth century, so much so that even in the 1640s a pamphlet entitled *A Short Essay Upon the Disorders of Scotland* could still speak of a 'defect of proportion in the Prince's authority'.[86] Seventeenth-century Scottish kings still had limited legislative power, no power to remove hereditary office-holders, no universal pardoning power, and an inability to hear some judicial cases on appeal. In certain sections of the country, moreover, such as in the Highlands and Islands, the king of Scots had no more power than that which he could maintain by force. By comparison the king of England, limited though he may have been by Continental standards, was a mighty potentate indeed. In 1648 an English pamphleteer, in opposing the peace proposals advanced by the Scottish commissioners, asked whether the Scots really wanted to give

[84] J. Bruce, *Report on the Events and Circumstances which produced the Union of the Kingdoms of England and Scotland* (London, 1799), 38.

[85] A. R. G. McMillan, *The Evolution of the Scottish Judiciary* (Edinburgh, 1941), 2–3; J. A. Lovat-Fraser, 'The Constitutional Position of the Scottish Monarch prior to the Union', *Law Quarterly Review*, 17 (1901), 252–7.

[86] 'A Short Essay upon the Disorders of Scotland', in *Miscellanea Aulica*, ed. T. Brown, 187–8.

their king ' the same power in Scotland, the same negative voice, the same absolute command and authority every way' that they claimed he had in England, especially since he might use this power to destroy presbyterianism.[87]

Although these restrictions on Scottish royal power did not disappear in the seventeenth century. Scottish kings developed ways of circumventing them and of creating the image, and in some respects the reality, of being a powerful, absolute monarch. This growth of royal power began with James VI, who not only developed a very high concept of kingship but actually gave some substance to it, even in the Highlands. James began this process of increasing royal power in the sixteenth century, but his great success came after, and to a great extent as a result of, the Union of the Crowns. The main reason for this is that once he was in England, James could run his northern kingdom without the threat of physical resistance which had always been a problem when he was in Edinburgh. 'Government by the pen', as he referred to it, was safer than personal government and it also proved to be more powerful.

A second reason for the increase in royal power after 1603 is that the acquisition of the English crown gave James the pretext to enlarge his power by statute.[88] After tightening his control of the Scottish parliament, so that it usually served as little more than a rubber stamp of royal policy. James and his successors secured the passage of a number of acts that made the king of Scotland, at least theoretically, as absolute as any prince in Europe.[89] One of these statutes was the Act anent the King's Prerogative of 1606, which was devised solely to give the king power that was commensurate with his recently enlarged imperial authority, and which recognized 'His Majesty's sovereign authority, princely power, royal prerogative, and privileges of his crown over all estates, persons and causes whatsoever within his said kingdom'.[90] Later in the

---

[87] *The Scottish Mist Dispel'd* (London, 1648), 31–2, 45.

[88] Dicey and Rait, *Thoughts on the Union*, 30.

[89] See Ridpath, *Discourse*, 50–4. On the parliamentary power of the king in the 16th and 17th centuries see G. Donaldson, *Scotland: James V to James VII* (Edinburgh, 1971), 279–87.

[90] *APS* iv. 281. In 1607 Bacon contrasted the indefiniteness of the English prerogative with the 'liberty' of the Scottish. *L&L* iii. 335–6.

century another Scottish statute, the Act of Supremacy of 1669, gave the king greater control over ecclesiastical affairs in Scotland than he had in England.[91] 'Never was king so absolute as you are in poor old Scotland', boasted Lauderdale to Charles II after the passage of that Act.[92] By the 1680s Alexander Mudie could write that 'The King [of Scotland] is an absolute and unaccountable monarch and (as the law calls him) a free prince of a sovereign people, having as great liberties and prerogatives by the laws of the realm, and privileges of his crown and diadem, as any other king or potentate whatsoever', an assessment with which Sir George Mackenzie could well agree.[93]

Both the traditional limitations on the power of Scottish kings and the attempts of the Stuarts to overcome or circumvent these limitations hampered union efforts in the seventeenth and early eighteenth centuries. In the early seventeenth century English anti-unionists had argued that the weakness of the Scottish king made union impossible or at least undesirable. As much as MPs may have disliked the absolutistic statements of King James, many of which had been formulated in the context of Scottish politics, they were much more concerned with the historic restrictions of his prerogative. They worried that the Scots 'make the estate of their king less monarchial than ours'.[94] In particular, they objected that the king of Scotland did not have a negative voice in the Scottish parliament. This objection was based on the widespread but by no means undisputed belief that acts passed by the Scottish parliament did not require the royal assent and that the tradition of touching the acts with the royal sceptre was a mere act of courtesy.[95]

It is uncertain whether English MPs raised this objection regarding the negative voice in order to counter plans for a parliamentary union or simply to demonstrate that the king could not control the Scottish nobility and that therefore union

---

[91] See Lee, *The Cabal*, 60–1; W. L. Mathieson, *Politics and Religion: A Study in Scottish History from the Reformation to the Revolution* (Glasgow, 1902), ii. 238.

[92] *The Lauderdale Papers*, ed. O. Airy (Camden Society, 1884–5), ii. 163–4.

[93] [Mudie] *Scotiae Indiculum*, 23; Sir G. Mackenzie., *The Institutions of the Law of Scotland* (Edinburgh, 1684), 17–18.

[94] SP 14/7/59.

[95] Lovat-Fraser, 'Constitutional Position', 252–3; Ridpath, *Discourse*, 167–8.

of any sort was dangerous. In any event, James felt obliged to show that in fact he never even allowed his Scottish parliament to consider matters without approval, and that if he did not agree with any measures which that assembly passed, he could always cleanse them from the record before ratification. 'If this may be called a negative voice', said James, 'then I have one, I am sure, in that Parliament.'[96] The main objective of James was to prove to his English parliament that there was 'nothing inclined to popularity' in the Scottish parliament. Popularity in Scotland had indeed been a main concern of English MPs. Sir Christopher Piggott was grossly exaggerating when he claimed that the Scots had not allowed more than two of their kings to die peaceably in their beds in the last two hundred years, but his statement reflected genuine English concern about the possible results of closer union.[97]

If the English used the weakness of the Scottish crown as an argument against union in the early seventeenth century, Scottish anti-unionists of the early eighteenth century used the revival of Scottish royal power since 1603 as one of their trump cards. Appealing to the traditional limitations of the Scottish king, the patriots of the early eighteenth century argued, quite correctly, that the increase in the Scottish prerogative during the past one hundred years had been the direct result of the Union of Crowns. George Ridpath, for example, claimed that James VI had invaded the freedom and constitution of the parliament of Scotland 'by his power and influence as King of England'.[98] More specifically, Ridpath, a presbyterian, attributed the growth of the royal prerogative in Scotland to the high church party in England, whose fear of presbyterianism had turned them against the imperfect union proposed in 1604. Seeking to distract James from this scheme, they had encouraged him instead to elevate his prerogative in Scotland.[99]

---

[96] *Political Works of James I*, 302.

[97] W. Cobbett, *The Parliamentary History of England from the Normans . . . to the Year 1803* (London, 1806), i. 1097. See also SP 14/7/59 for the argument that union may be dangerous to the king's person.

[98] [G. Ridpath], *The Reducing of Scotland by Arms and Annexing it to England as a Province Considered* (London, n.d.), 19–20.

[99] Ridpath, *Discourse*, 44. Fletcher made the same argument, although he was more critical of the English presbyterians. Daiches, *Scotland and the Union*, 73.

As far as Ridpath and his associates were concerned, the events of the seventeenth century had only made the situation worse. Even the Glorious Revolution had not provided relief, since the king could still exercise considerable power over Scottish affairs through the Scottish privy council, without submitting his policies to parliamentary review. His power, moreover, already had a statutory foundation, so he was on solid legal ground acting in this manner.[100] To make matters worse, the king could thwart any Scottish parliamentary initiatives that were hostile to English interests. One could argue, as did Andrew Fletcher of Saltoun, that the problem was not the prerogative of the Scottish king but the prerogatives of English ministers over Scotland, but to most Scots the problem was the Scottish royal prerogative, as nourished in England after the Union of the Crowns.[101]

For the Scots, the solution to this problem of royal power was either complete independence, strict limitation of the king's Scottish prerogative, or a federal union. The solution that was adopted, incorporating union, had little appeal, since it only promised to make the lethal English connection stronger. Nevertheless, the incorporation of the Scottish parliament, which despite the events of 1690 was not a sovereign assembly,[102] into a sovereign English parliament that could enforce restrictions on royal power, coupled with the *de facto* control over Scottish affairs entrusted to Scottish MPs, did go a long way towards resolving the issue. Scotland may have been unhappy about its constitutional position after 1707, but it could no longer claim that the king exercised unlimited power in the country by virtue of his Scottish prerogative.

IV

The third set of constitutional problems connected with the union concerned the executive—the councils and administrative institutions through which the king ruled both countries. The Union of the Crowns had raised two separate but related

---

[100] W. S. McKechnie, 'The Constitutional Necessity for the Union of 1707', *SHR* 5 (1908), 58.

[101] Ibid. 61.

[102] Dicey and Rait, *Thoughts on the Union*, 241.

questions. First, would Scots and Englishmen be capable of holding offices, ecclesiastical as well as temporal, in the other country? Second, would any of the councils, courts, or administrative departments of either country be remodelled so that they would have jurisdiction over both countries? The first question was of greater concern in the early seventeenth century, the second in the early eighteenth century.

Scottish participation in English offices was perhaps the most heated political issue associated with the union at the beginning of the seventeenth century. The arrival of a number of Scots in London at the time of the Union of the Crowns, coupled with the possibility of the naturalization of all Scots in England, led many Englishmen to fear that England would be overrun with 'swarms of tawny Scots, who would rule all'.[103] The prospect of Scottish appointments was especially unwelcome at this time, since the English government was already having difficulty providing administrative positions for all those Englishmen who desired them. The appointment of Scots, therefore, threatened to make a bad situation even worse. Nevertheless, even the most severe critics of the king were not prepared to deny his right to make whatever appointments he wished. All they could do was to ask him not to appoint any Scots to judicial positions on the grounds that Scots did not know English law. Sir John Bennet, an English ecclesiastical judge, made such a proposal in 1604, and in similar fashion Sir George More, a common lawyer, recommended that the prohibition be extended to include the commissioners of the peace and coroners.[104]

James proceeded with great caution on this matter of appointments. In a letter to the commissioners in 1604 he agreed not to appoint Scots to any office of judicatory or place in parliament, even though his prerogative entitled him to do so.[105] Aside from the appointment of Edward Bruce, Lord Kinloss, as Master of the Rolls, which he made almost immediately upon his arrival in England and before English

[103] *CJ* i. 359.
[104] Ibid. 362. A somewhat different proposal of that year suggested that the four main positions in government would always be filled by Englishmen and that in all other appointments there would be a twelve-year moratorium on the appointment of Scots. *CSP Ven. 1603–7*, 48.
[105] SP 14/10A/40 I. See also Defoe, *History of the Union*, 722–4.

sentiment on this question had been expressed, he kept his promise. With respect to non-judicial appointments James also exercised considerable restraint. With the exception of George Home, the Earl of Dunbar, who served as the Chancellor of the Exchequer from 1603 to 1605, and three other Scots who were given relatively minor administrative posts, James excluded Scots from the departments of state, confining them instead to the royal household.[106] The necessity of having a household for both king and queen for the first time in more than fifty years, as well as one for Prince Henry, made possible the absorption of a substantial number of Scots—158 in all—into the king's government without displacing many Englishmen.[107] James also made a few appointments of Scots to positions in local government, mainly in the North, where Scottish peers often held land. The most important of these appointments was the naming of the Earl of Dunbar to the lieutenancy of Northumberland, Cumberland, Westmorland, and Newcastle in 1607.[108]

Although James named very few Scots to English administrative posts, he did not hesitate to add a number of Scots to the English privy council, a practice that his son continued.[109] The number of Scots on the council fluctuated from year to year, reaching a high of seven on a number of occasions during the reign of Charles I. In 1633 Charles named a comparable number of Englishmen to the Scottish council.[110] Neither king, however, ever tried to appoint a single council for all of Britain, as Sir Francis Bacon, David Hume, and one anonymous union writer had all recommended in 1604.[111]

[106] D. H. Willson, 'King James I and Anglo-Scottish Unity', in *Conflict in Stuart England: Essays in Honour of Wallace Notestein*, ed. W. A. Aiken and B. D. Henning (London, 1960), 47–8; P. R. Seddon, 'Patronage and Officers in the Reign of James I', Ph.D. thesis (Manchester, 1967), 166.

[107] Seddon, 'Patronage', 161, 293–305. This number includes all the Scots who received appointments in England. Of these all but nine held at least one position in the household.

[108] S. J. Watts, *From Border to Middle Shire: Northumberland 1586–1625* (Leicester, 1975), 152. Sir Thomas Kilpatrick was appointed as Surveyor of lands in the North and Lancaster, while he and Adam Newton both served as secretaries to the Council in the Marches of Wales. Seddon, 'Patronage', 166.   [109] *CSP Ven. 1603–7*, 33.

[110] G. E. Aylmer, *The King's Servants: The Civil Service of Charles I* (London, 1961), 24–5.

[111] *L&L* iii. 229; Hume, 'Tractatus Secundus', NLS, Advocates MS 31.6.12, ff. 50ᵛ–51; 'Pro Unione', Gonville & Caius College, Cambridge, MS 73/40, f. 188.

The mutual participation of Scots and Englishmen in the offices of the other country was not a union scheme as such, since it did not in any way encourage a structural unification of the two governments. In a certain sense it actually worked against the prospect of further union, since it enabled the king to co-ordinate the governance of his two kingdoms without fully uniting them. Nevertheless, the unrestricted participation of Scots in English offices, which came about only after the Treaty of 1707, can be considered one of the attractions that union had in the northern kingdom. Just as union offered Scottish merchants full participation in English overseas trade, so it offered Scottish administrators full participation in the burgeoning civil service. In some departments of government, such as the colonial service, Scots filled positions in numbers far out of proportion to the percentage of Scots in the British population.[112] In other departments, especially in the judiciary, they were less numerous and conspicuous, but they nevertheless managed to make their mark. Within fifty years of the union of 1707 Scotland had produced not only a Lord Chief Justice, Mansfield, but a Prime Minister, Bute.

The participation of Scots and Englishmen in the offices of the other country was just one side of the administrative union question; the actual fusion of the bureaucratic institutions of the two countries was the other. In the early seventeenth century very little attention was given to the latter question. Sir Francis Bacon recommended the establishment of a court, similar to the Grand Council of France, which could draw cases from the ordinary jurisdictions of both countries, as well as a court at Berwick to handle matters that concerned both the English and the Scottish Borders.[113] Richard Hudson hoped that King James would include not only Englishmen and Scots but also some Irishmen in a 'Grand Council of State'.[114] Another anonymous writer, while not actually calling for the creation of new, joint institutions, proposed occasional joint sessions of the English and Scottish councils,

---

[112] I. C. C. Graham, *Colonists from Scotland: Emigration to North America, 1707–1783* (Ithaca, NY, 1956), 17; S. S. Webb, *The Governors-General: The English Army and the Definition of Empire* (Chapel Hill, NC, 1979), 444–5.

[113] *L&L* iii. 221, 223.

[114] *Calendar of the State Papers relating to Ireland of the Reign of James I, 1603–1606* (London, 1872), 238.

a plan that anticipated the later parliamentary proposal of Lauderdale.[115] None of these plans found much support in the early seventeenth century, so with the exception of the joint commissions that James appointed to maintain order in the Borders, English and Scottish administration remained completely independent of each other.

It was not until the time of the Civil War that any new efforts were made to govern England and Scotland through some sort of general, Anglo-Scottish institutions. The first such proposal came from the Scottish commissioners as part of their eighth demand in 1641. In order to preserve the peace between the two kingdoms they suggested that a certain number of Scotsmen 'be placed about the King and Queen and the Prince'.[116] This proposal formed part of a general plan to create a British court, 'a supranational institution, composed of individuals from each of the two kingdoms, but itself belonging to neither of them'.[117]

Although the Scottish proposal was never adopted, a very different type of supranational institution was established three years later. In 1644 the necessity of co-ordinating the English and Scottish war efforts led to the formation of the Committee of Both Kingdoms, which consisted of seven English peers and fourteen MPs who were instructed to act jointly with the four Scottish commissioners in London.[118] They were instructed to 'advise, consult, order and direct concerning the carrying on and managing of the war' and 'to hold good correspondence and intelligence with foreign states'.[119] Since England and Scotland were at this time joined merely in a 'Solemn League and Covenant' rather than in an incorporating union, and since the personal union had by this time been reduced to a constitutional technicality, this

[115] 'The Devine Providence', Beaulieu Palace House, Papers on Scotch Affairs, III, item 1.

[116] BL, Stowe MS 187, f. 46. For the king's reply see J. Rushworth, *Historical Collections* (London, 1721), ii. 368. In 1646 the commissioners repeated the demand, this time calling specifically for one half, or at least one third of all positions of trust to go to Scots. SRO, PA 7/24, f. 206.

[117] W. Makey, *The Church of the Covenant, 1637–1651: Revolution and Social Change in Scotland* (Edinburgh, 1979), 65.

[118] W. Notestein, 'The Establishment of the Committee of Both Kingdoms', *AHR* 17 (1912), 481–2.

[119] *Constitutional Documents of the Puritan Revolution*, ed. Gardiner, 272.

Committee achieved an astonishing degree of administrative co-ordination. More than an *ad hoc* local commission, such as James I's Border commission, it exercised genuine executive functions, although it maintained a dependence upon the parliaments of both countries. Even if it was primarily an English institution,[120] similar in national representation to the English privy council under Charles I, it constituted the first real institutional tie between the two nations.

But not the last. Soon after the disbandment of the Committee, Scotland was conquered by, and incorporated into England, a process that entailed the destruction not only of the Scottish parliament but the Scottish privy council as well. The functions previously exercised by the council were assumed by the Council of State, which in turn delegated the administration of the northern kingdom to the commissioners for the administration of justice and the army.[121] Scotland, therefore, for the first time since the Union of the Crowns, was governed by distinctly Anglo-Scottish governmental agencies, staffed by both Scots and Englishmen. As it turned out, however, these institutional arrangements were only temporary, and by the 1660s an older pattern had re-emerged, in which the king governed his two kingdoms through their own native institutions, occasionally employing men like the Duke of Lauderdale on the councils of both countries in order to co-ordinate them.[122]

When the possibility of an incorporating union arose again in the late seventeenth and early eighteenth centuries, the problem of governing Scotland once again came to the fore. Somewhat surprisingly, neither the union commissioners nor the authors of union pamphlets gave much attention to this question, and the Union Treaty of 1707 left many aspects of it unresolved. In the most general terms the question was whether to create new British institutions, which usually meant extending English institutions to Scotland, or to establish (or preserve) distinctly Scottish institutions within the union. In almost all cases the latter alternative was adopted.

[120] Ferguson, *Scotland's Relations with England*, 127.

[121] A. R. G. McMillan, 'The Judicial System of the Commonwealth of Scotland', *JR* 49 (1937), 232–55.

[122] Donaldson, *Scotland: James V–James VII*, 290.

The only British institutions that were established by the Union were parliament, the largely honorific Lord Chancellorship, and the Treasury. It is true of course that the Treaty of 1707 laid the basis for the destruction of the Scottish privy council, which occurred in 1708, but instead of absorbing it into the existing English institution, parliament replaced it with two Scottish institutions, the Commission of Chamberlainy and Trade and the Commission of Police. In 1727 the economic functions of the council were transferred to another Scottish institution, the Board of Manufactures.[123]

In dealing with financial matters the union brought yet another solution to the problem of Scottish administration. Since the Treaty insisted that the revenue and customs systems would be the same throughout the United Kingdom, there was no possibility that Scotland could be given much autonomy in this regard. The abolition of the Scottish Treasury, therefore, and the decision to give final control of financial matters to the Treasury of the United Kingdom were foregone conclusions. But the actual administration of Scottish finance was none the less granted to distinctly Scottish boards of customs and excise (which were merged with the English boards only during the period 1723–42) and to a Scottish Court of Exchequer which, while being constructed on an English model, was entirely separate from its English counterpart.[124]

All these administrative changes suggest that there was never any intention on the part of Englishmen simply to assimilate Scotland to England in 1707, or, for that matter, at any time thereafter. Scotland was never treated as 'Scotlandshire', a northern and remote land similar to Westmorland or Northumberland. Indeed, even the individual shires of Scotland were not treated in the same way as the northern shires of England, since the Scottish system of local administration was different from the English system in 1707 and has remained that way ever since. The fact of the matter is that for very practical reasons Scotland could not be governed as just another part of England. Having existed for centuries as a

---

[123] A. L. Murray, 'Administration and the Law', *The Union of 1707: Its Impact on Scotland*, ed. T. I. Rae (Glasgow, 1974), 32.

[124] Ibid. 36.

separate kingdom, it could not be readily absorbed into an expanded version of the English state. Its parliament was, to be sure, completely absorbed into the English parliament, but in passing legislation that affected Scotland that assembly very often decided to treat the northern kingdom separately from the rest of Britain. And in certain areas of administration, not to mention its completely separate judiciary, Scotland continued to function as if it were still an autonomous kingdom.[125]

The separate administration of Scotland after the union raises some important questions about the nature of the British state. The first is whether the Scottish state continued to exist in some form or other after the union, since certain aspects of the old Scottish constitution persisted after 1707.[126] The problem with this line of questioning is that virtually all definitions of the state include the possession of sovereignty as one of its essential attributes, and it is clear that Scotland surrendered its sovereignty in 1707.[127] It is much more useful to ask, in light of the administrative structure of the British state, whether that state was unitary. The conventional wisdom is that it was, since the Scottish parliament was incorporated into that of England. But parliament was only one of many institutions that exercised political power within the new British state. Once the entire range of British governmental institutions is examined, the British state loses much of its unitary character.

If the political terms of the union settlement call the unitary character of the British state into question, then its legal and ecclesiastical features make it appear to be unequivocally pluralistic. We must remember that in the seventeenth century Englishmen and Scots considered both the legal

[125] For a discussion of the 'administrative quasi-federalism' of Scotland in the 20th century see W. J. M. Mackenzie, 'Peripheries and Nation-Building: The Case of Scotland', in *Mobilization, Center-Periphery Structures and Nation-Building*, ed. P. Torsvik (Bergen, 1981), 165–6. See also R. K. Murray, 'Devolution in the U.K.: A Scottish Perspective', *Law Quarterly Review*, 96 (1980), 41.

[126] See R. Sutherland, 'Aspects of the Scottish Constitution Prior to 1707' in *Independence and Devolution: The Legal Implications for Scotland*, ed. J. P. Grant (Edinburgh, 1976), 18–21.

[127] The argument of Francis Grant that the Treaty would not deprive Scotland of her sovereignty since 'a main part of it remains amongst ourselves' presupposes a very different view of sovereignty. *The Patriot Resolved*, 10.

system and the system of church government to be essential features of the body politic or, as we would say, the state. In the British state constructed in 1707, however, there was a deliberate effort made to keep both English and Scots law and the English and Scottish churches separate. Why these decisions were made will be the subject of the next two chapters.

3

The Union of Laws

THE pamphleteers, politicians, and other public figures who argued for and against the union in the early eighteenth century do not appear to have been greatly concerned with the relationship between English and Scots law. The main reason for this lack of contemporary interest is that after 1670 most English and Scottish unionists agreed that any treaty of union would preserve the integrity and independence of Scottish private law and would leave even a large part of the public law of the two kingdoms distinct. Since this agreement was generally regarded as a concession to those who wished to preserve as much Scottish autonomy as possible within the union, it did not attract a large amount of criticism from anti-unionists, either in the Scottish parliament or in the Scottish press. In neither country did opposition to the Treaty of Union focus on its legal provisions, and when discussion of the law did arise, it dealt with those questions that the Treaty ignored or postponed rather than with those that it addressed.[1]

The fact that the governments of both countries reached an understanding regarding the legal aspects of the union should not, however, minimize the significance of that agreement. The acceptance of a legal pluralism within a politically unitary state ran counter to the assumptions which Englishmen and Scots had made for centuries regarding the relationship between the law and the state. Nor should the agreement that was embodied in the Treaty of 1707 detract from the importance that both Englishmen and Scots had attached to the question throughout the seventeenth century.

[1] [G. Mackenzie, Earl of Cromarty], *A Letter to a Member of Parliament upon the 19th Article of the Treaty of Union between the two Kingdoms of Scotland and England* (n.p., 1706); [J. Spottiswoode], *The Trimmer: Or, Some Necessary Cautions Concerning the Union of the Kingdoms of Scotland and England* (Edinburgh, 1706); *Overture for an Additional Clause to the Nineteenth Article, anent the Session in Scotland* [n.p., 1706].

The decision made in the early eighteenth century to accept a qualified legal pluralism within the British state was reached only after numerous proposals for legal union had been advanced and debated. These proposals held such great appeal to unionists on both sides of the Tweed that even after 1670, when it appeared that the issue had for all practical purposes been settled,[2] new schemes for uniting the laws continued to surface.[3] The dream of legal union died slowly. In the middle of the eighteenth century Lord Hardwicke still regretted that the great bulk of English and Scots law had not been amalgamated in 1707, while Lord Kames hoped that such a union might still be achieved.[4]

This chapter will study the various proposals for legal union that appeared at different times throughout the seventeenth century, especially between 1603 and 1670. It will explain why ideas of legal union appealed so strongly to many individuals on both sides of the Tweed and why the same ideas encountered determined and often implacable opposition. It will also examine the practical difficulties inherent in any such scheme and assess the significance of the legal provisions of the Treaty of 1707.

I

When Englishmen and Scots contemplated the possibility of strengthening the Union of the Crowns, they often thought in terms of legal union. If the union were to be an incorporating one, a union of the political communities of the two kingdoms, then a union of laws appeared to be a necessity. Laws were, in contemporary political thinking, the sinews of the body politic. From this it followed that if there was to be a new British body politic, then there would have to be some sort of

---

[2] On the commitment of both the government and the Scottish commissioners to legal pluralism at this time see SP 104/76/157, art. 10; G. M. Hutton, 'Stair's Public Career', in *Stair Tercentenary Studies*, ed. D. Walker (Stair Society, 1981), 11–12.

[3] [Fairfax], *A Discourse*; [W. Seton], *The Interest of Scotland in Three Essays* (2nd edn., Edinburgh, 1702), 47; [Chamberlen], *The Great Advantages*, 20; [J. Buchan], *A Memorial Pointing to Some of the Advantages of the Union of the Two Kingdoms* (London, 1702), 7.

[4] Trevor-Roper, *Religion, the Reformation and Social Change*, 467; J. W. Cairns, 'Institutional Writings in Scotland Reconsidered', in *New Perspectives in Scottish Legal History*, ed. A. Kiralfy and H. L. MacQueen (London, 1984), 101.

legal union. The assumption was so instinctive that John
Hayward entitled one chapter of his pro-union treatise 'Of the
Bodily Union, or Union by Law'.[5] John Thornborough, like
Hayward, argued that if the body was to be one with the head,
then there would have to be a union of laws and love.[6] Even
the English judges, in opposing the renaming of the two
kingdoms as 'Great Britain' in 1604, revealed how closely
Englishmen associated political and legal union. As far as the
judges were concerned, a new kingdom of Great Britain,
which they believed the adoption of a new name would
signify, could not be erected until there was 'an union of the
laws of both kingdoms'.[7]

The tendency to think in terms of legal union did not
derive simply from the logic of contemporary political
thought. The historical experience of England and to a lesser
extent Scotland also made the idea of legal union a natural
corollary to that of political union. Both kingdoms, themselves
composites of smaller kingdoms and principalities, could be
considered legally unified. Both possessed a common law and
a central, law-making parliament. In England, which had the
oldest national law in Europe, legal unity was in fact the most
salient feature of the state. In Scotland, where the law was
much less developed and systematic than in England, where
parliamentary statutes had much less authority, and where
custom would often conflict with law, there was still much
greater legal cohesion than in other European states.
Certainly the existence of a common law in each country
conditioned men to think in terms of British legal unity as
either a prerequisite or a necessary concomitant of a union of
the kingdoms.

A more practical consideration that led both Scots and
Englishmen to discuss the possibility of legal union was the
need for the king to administer justice uniformly throughout
his recently enlarged dominions, especially in the notoriously
lawless area of the Borders. It was this need that inspired Sir
Francis Bacon to propose as early as 1604 the establishment of

---

[5] Hayward, *Treatise*, 8. See also Craig, *De Unione*, 328.

[6] Thornborough, *Discourse*, 17. See also id., *Ioiefull Reuniting*, 77.

[7] Coke, *Fourth Part of the Institutes*, 347. Ellesmere also recognized that further union
would require a union of laws. See 'Speech touching the *Post-Nati*', in Knafla, *Law and
Politics*, 251.

an Anglo-Scottish court at Berwick administering English law, Scots law, or a mixture of the two.[8] The same need explains why both parliaments, in dealing with proposals for the remanding of prisoners across the Border for trial in the country where they had committed their crimes, recognized the benefits of judicial or legal unity. The Scottish parliament made the emphatic point that measures to prevent crimes in the neighbouring country and to punish the malefactors 'can not goodly be unless the justice of both realms in such causes be uniform and no material difference be in the substance or order of their judicatories in actions of that kind',[9] while English MPs, in arguing against remanding, remained open to the possibility of a 'treating of the conformity of laws of both kingdoms'.[10]

The mere prospect of a politically and administratively united Britain, therefore, led Englishmen and Scots to consider the possibility of legal union. When these men referred to a 'union of laws', however, they did not always mean the same thing. The phrase was used to denote different types of legal union as well as the union of different areas of law. With respect to the type of union there were three general possibilities: conformity, uniformity, and fusion. The conformity of laws meant that the two laws would stand in general agreement, while uniformity, a more exacting standard, meant that they would be essentially the same, if not identical. The terms were in fact often used interchangeably. The achievement of either conformity or uniformity did not require the prior establishment of a united law-making assembly. Specially appointed commissions of lawyers from both nations, acting with the support of both parliaments, were capable of achieving the desired correspondence between the laws. The third possibility, however, legal fusion, which meant that the same body of law would apply to both countries, did require the formation of a single British parliament.

Some of the most ardent advocates of legal union, especially in the early seventeenth century, favoured nothing more than

---

[8] *L&L* iii. 221.
[9] *APS* iv. 367.
[10] SP 14/27/30.

the achievement of conformity or uniformity. King James VI and I, for example, expressed the hope in 1604 that by the time he died there would be established in Britain 'one worship to God, one kingdom entirely governed, one uniformity in laws'.[11] Three years later, when he told the English parliament that he hoped one day they would 'all be governed by one law', he meant no more than the uniformity he had referred to in 1604.[12] Since by that time James had abandoned the hope of establishing a common British parliament, at least during his lifetime, uniformity was all he could expect. And in fact, from the point of view of 'government', uniformity was all he needed. It made little difference to him whether he governed his two kingdoms through one common set of laws or through two that were for all practical purposes the same.

Among all the plans that were advanced for a union of laws between 1603 and 1707, very few actually called for their fusion. Hayward apparently had such a process in mind when he recommended the reduction of the 'laws of England and Scotland into one body', but his plan actually dealt more with the reconciliation of differences and the recognition that certain laws were already in agreement than with the creation of one common code.[13] A similar plan, according to which an equal number of commissioners from both countries would reduce English and Scots law into a system for 'all Great Britain' was drafted in 1669 as an alternative to the scheme submitted to the union commissioners in 1670.[14] Two early eighteenth-century legal unionists, William Seton of Pitmedden and Blackerby Fairfax, also advocated the formation of a common body of British law, but this was to be accomplished only after the Treaty of Union was ratified.[15] Among the advocates of legal fusion one might also wish to consider the English legal imperialists, such as the anonymous author of *A*

[11] *CJ* i. 171. In dealing with remanding parliament claimed that its objective was to 'reduce our laws to uniformity'. *Bowyer Diary*, 316 n.

[12] *Political Works of James I*, 292. See also SP 15/36/66. In 1616 James told his English subjects that he wished to 'conform the laws of Scotland to the laws of England', *Political Works*, 329. The question of what James meant by legal union is still uncertain. See T. B. Smith, 'British Justice: A Jacobean Phantasma', *The Scots Law Times*, 4 June 1982, 157–64.

[13] Hayward, *Treatise*, 14.

[14] NLS, MS 597, f. 234.

[15] Seton, *Interest*, 47; Fairfax, *Discourse*, 52–3.

Discourse upon the Union (1664), who called for Scotland to abandon its laws and become legally assimilated to England.[16] In such circumstances legal fusion would be achieved simply by imposing English law on Scotland, as it had been imposed on Wales and Ireland.[17]

The phrase 'legal union' as well as embracing plans for the conformity, uniformity, and amalgamation of the two laws, also covered plans for uniting different areas of law. The great majority of legal unionists insisted that they wished to achieve a union of only public, not private law. The distinction between these two types of law derived from civil law, but it could be applied, with some difficulty, to both English and Scots law.[18] Public law included those laws that dealt with 'public right, policy, and civil government', while private law encompassed those laws that concerned *meum et tuum* or private right.[19] Public law comprised not only many aspects of fundamental law—the law of the constitution—but all 'laws of government', including the criminal law. Much, although not all, of it was embodied in statute, and therefore a rough correlation prevailed between *lex* and public law on the one hand and *jus* and private law on the other. The terms cannot be equated, however, since a substantial amount of public law, especially fundamental law, was contained both in custom and case law, while statutes often made significant alterations in private law. Perhaps the best equivalents of public and private law are 'laws of state' and 'laws of controversy', terms which were used in one early seventeenth-century union treatise.[20]

There are three main reasons why most plans for legal union dealt only with 'laws of state' or public law. First, the

[16] 'A Discourse upon the Union', in *Miscellanea Aulica*, ed. Brown, 196.

[17] For the extension of the common law to Ireland see H. S. Pawlisch, *Sir John Davies and the Conquest of Ireland: A Study in Legal Imperialism* (Cambridge, 1985); *Hibernica or Some Ancient Pieces relating to Ireland*, ed. W. Harris, Pt. II (Dublin, 1750), 33, 128.

[18] For English applications see Sir F. Bacon, 'A Preparation toward the Union of the Laws of England and Scotland', in *Works*, ed. B. Montague (Philadelphia, 1855), ii. 161; the statement of Sir Daniel Dun, *CJ* i. 1020; R. Zouch, *Elementa Jurisprudentiae* (Oxford, 1636), 28. On the failure of Teutonic and common law to make the distinction see H. F. Jolowicz, 'Political Implications of Roman Law', *Tulane Law Review*, 22 (1947), 61–81. See generally D. Walker, *The Scottish Legal System* (Edinburgh, 1981), 78–80.   [19] Pryde, *Treaty of Union*, 95.   [20] SP 14/7/61.

main benefit that legal union promised was the uniform administration of justice throughout the two kingdoms, especially in the Borders, and the realization of this goal required only a union of the public, and in particular the criminal laws of both countries. It is not surprising that Bacon, in outlining his programme for legal union, proposed that attention first be given to the criminal statutes.[21] Second, changes in private law, the law by which men held and disposed of their property, were difficult and even dangerous to undertake, since 'men love to hold their own as they have held'.[22] The reluctance of English unionists to tamper with the private law of Scotland became most apparent in the 1650s, when the English parliament had the opportunity to change as much of Scots law as it desired. It specifically enjoined its designated commissioners, however, 'to see that the laws of England *as to matter of government* be put in execution in Scotland'.[23] Third, Englishmen were aware that the preservation of numerable local customs, most of which dealt with private law, could easily be accommodated within a national law. If the particular customs of Kent and Wales had been preserved within the broad framework of the English common law, then the 'particular privileges' and peculiar 'municipal customs' of Scotland could likewise be preserved within the framework of a some future body of British law.

Legal unionists may have displayed a good deal of common sense in restricting most of their proposals to the area of public law, but nevertheless they sometimes envisaged a more comprehensive legal union. Even Bacon, who was less susceptible to theoretical flights of fancy on this issue than other legal unionists, foresaw the eventual Scottish acceptance of English customs, which would of course be reflected in their private law.[24] Sir Thomas Craig, in arguing that English and

---

[21] Bacon, 'A Preparation', in *Works*, ii. 161. Between 1607 and 1610 Thomas Ashe did compile a 'Brief Collection of All General Penal Statutes', House of Lords Record Office, Braye MS 50, ff. 5–33. Ashe's work may have been inspired by the union.

[22] Bacon, *Works*, ii. 160.

[23] *Cromwellian Union*, p. xix. Emphasis added. Justice in Scotland during the interregnum was administered according to the laws of Scotland and any laws made by parliament. See Add. MS 4158, ff. 102ᵛ–103; Edinburgh University Library, Laing MS I. 290.

[24] *L&L* iii. 335–6. See [D. Defoe], *Remarks upon the Lord Havarsham's Speech in the*

Scots law were so similar that they *could* be united, even if that were inadvisable, dealt with private as well as public law.[25] When John Cowell prepared a digest of English law in 1605 in order to encourage some sort of legal union, he included much private law in his work.[26] A half century later Sir Thomas Urquhart, a Scottish supporter of Cromwell who was very much taken with the idea of legal union, called for 'an identity of privileges, laws and customs', between England and Scotland, not simply a union of public law.[27] The few plans for legal union formulated after the Restoration also included provisions for the union of private as well as public law.[28]

Most of the proposals for legal union concerned the substantive law, either public or private, of the two countries, but a few dealt with the legal procedures used in their courts and with the courts of law themselves. The only time that procedural differences became the subject of debate in connection with the union was during the consideration of remanding in 1607; at other times unionists gave only passing attention to such matters. There were also very few proposals for uniting the judicial institutions of the two countries. As mentioned above, Bacon proposed the erection of an Anglo-Scottish court at Berwick that could administer either English law, or Scots law, or a mixture of the two, but the idea attracted no apparent support. During the Cromwellian period English judges joined Scottish judges in trying Scottish cases, but the courts themselves remained essentially Scottish, not Anglo-Scottish institutions. After that time the only serious proposal for judicial union appeared in 1706. It envisaged the erection of an Anglo-Scottish court of Exchequer at Berwick to deal with the legal problems of the commercial union that the Treaty of 1707 brought about. This hybrid court was viewed as a remedy for a specific set of problems that would attend the implementation of the Treaty, not as a

House of Peers, February 15, 1707 [Edinburgh, 1707], for an early 18th century interpretation of Bacon's ideas on this point.

[25] Craig, *De Unione*, 304–20.

[26] J. Cowell, *Institutiones Juris Anglicani* (Cambridge, 1605).

[27] Sir T. Urquhart, *Tracts of the Learned and Celebrated Antiquarian Sir Thomas Urquhart of Cromarty* (Edinburgh, 1774), 3, 163.

[28] 'Discourse upon the Union', *Miscellanea Aulica*, 194; NLS, MS 597, f. 234; Seton, *Interest*, 47; Fairfax, *Discourse*, 51; Chamberlen, *Great Advantages*, 20.

pilot project for a more thorough union of the judicatories of the two nations.[29]

II

Proposals for Anglo-Scottish legal union acquired much of their strength from the close similarities that English and Scots law exhibited. In the proclamation of 1604 by which he assumed the style of King of Great Britain, James I summarized those similarities. He declared that he had learned from those skilled in the laws of England that

> there is a greater affinity and concurrence between most of the ancient laws of both kingdoms than is to be found between those of any other two nations, as namely, in states of inheritance and freehold, as fee simple, fee tail, tenant for life, by courtesy, dower and such like; in cases of descents of inheritance; in tenure of lands, as of knights service, socage, frankalmoign, burgage, villeinage and such like; in writs and forms of process; in cases of trial by juries, grand juries; and lastly in officers and ministers of justice, as sheriffs, coroners, and such like . . .[30]

It is not certain from which lawyers James derived these points of comparison. Whoever they were, they did not have a thorough knowledge of Scots law, since Scotland did not have a grand jury. The lawyers were also not necessarily ardent unionists, for Coke, who was generally negative on the question of Anglo-Scottish unity, listed many of the same similarities in his *Institutes*.[31] It was certainly possible for one to do so in the spirit of detached scholarly interest. Nevertheless, the basic similarities between the laws of the two countries became the stock-in-trade of unionists throughout the seventeenth century.[32] Often unionists used these comparisons to prove that the people of both countries were prepared for social union; similar laws, just like a common language and a common religion, ensured that union would be

---

[29] *An Essay upon the Union of the Kingdoms of England and Scotland* (London, 1705), 8.

[30] *Stuart Royal Proclamations*, i. 95–6.

[31] Coke, *Fourth Part of the Institutes*, 347. See Sir J. Eliot, *De Jure Majestate*, ed. A. B. Grosart (London, 1882), 163 for a similar comparison.

[32] See, for example, Thornborough, *Reuniting*, 9; Robert Pont, 'Of the Union of Britayne', in *Jacobean Union*, 24; Bacon, 'Preparation' in *Works*, ii. 161; BL, Sloane MS 1786, ff. 101–2; BL, Cotton MS Titus F IV, f. 63ᵛ.

commodious rather than traumatic. At other times, however, the similarities that prevailed between English and Scots law were used directly to support the case for legal union.

The unionist who developed these points of comparison most fully was the Scottish lawyer, Sir Thomas Craig. His motives for doing so are difficult to determine. To some extent he was engaging in an academic enterprise. In addition to being a practising lawyer, Craig was a legal scholar who had studied in France and had acquired the historical outlook of the great sixteenth-century legal humanists Budé, Cujas, and Hotman.[33] Using the methods of this school he studied the vast bulk of feudal law. In *Jus Feudale*, written in 1603, Craig showed that the Normans had introduced basically the same feudal law into England and Scotland, a claim that challenged the assumption of English common lawyers that their law pre-dated the Conquest and had never changed. Two years later, when Craig wrote *De Unione Regnorum Britanniae*, he drew heavily upon his earlier work to illustrate the similarities between the two legal systems.

The sections of *De Unione* that dealt with English and Scots law were therefore the product of mature scholarly research. Craig had not written *De Unione*, however, as a scholarly treatise. As one of the most committed of the Scottish unionists, and as one of the commissioners for the union in 1604, Craig had written the book both as a plea for further union and as an apology for the commission. Perhaps because the commission had not formulated a plan for legal union, Craig felt it was incumbent upon him to defend the work of the panel on which he had served.[34] He argued strongly that the perfection of the union did not require the laws of the two countries to be identical, and he recommended in the strongest possible terms that 'each nation be governed in accordance with its own laws and customs'.[35]

If this statement accurately represents Craig's attitude regarding legal union, then the question remains why he devoted such a large portion of his treatise to demonstrating

---

[33] J. G. A. Pocock, *The Ancient Constitution and the Feudal Law*, (Cambridge, 1957) 72–90.

[34] Craig, *De Unione*, 275.

[35] Ibid, 303.

the similarity of the two laws. Such detail was not necessary in order to prove that the laws were sufficiently in agreement to prevent the dissolution of a 'contract of lasting friendship' between the two countries.[36] The only reasonable interpretation is that Craig was hoping to lay the foundation of legal union in the future. Indeed, he concluded his chapter on the laws of England and Scotland with a clear statement to that effect. 'Nor is there any reason', wrote Craig, 'to despair of the possibility of so harmonising the legal systems of the two peoples as to fashion one body of law applicable equally to both, and thereby to promote the union of the two countries in one body politic.'[37] Craig was very much taken with the idea of legal union, but he recognized that its achievement in the early seventeenth century would be inadvisable. Doubtless he agreed with Bacon and other unionists that legal change was dangerous and should not be attempted rashly. As a staunch defender of Scottish sovereignty he also recognized the possibility that the early implementation of legal union might result in the violation of Scotland's constitutional integrity. He decided, therefore, to oppose any immediate progress on the question but still leave the door open for legal union in the future.

But how was such a union to be achieved? As much as Craig might celebrate the common origin of the two legal systems, he could not deny that they had diverged considerably once the countries had become independent of each other. Somewhat unrealistically, Craig recommended that the two nations seek a foundation of their legal union in the original Norman code or, failing that, in the original feudal law from which the Norman code had been derived. If this union were not to work (and who could honestly have expected both nations to have forsaken centuries of legal development in search of recovering an ancient unity?), then it was necessary to 'go to the Civil Law, whose principles are so equitable and of such widespread acceptance that it deservedly merits the appellation common law'.[38]

At first blush this recommendation strikes one as totally

---

[36] Ibid. 465.
[37] Ibid. 328.
[38] Ibid. 327–8.

absurd. English law in the early seventeenth century could be characterized mainly by its distinction from the civil law. One of the primary arguments against legal union, as shall be discussed below, was that English law could not be fused with Scots law precisely because the latter had been subject to civilian influences. In proposing the civil law as the foundation of Anglo-Scottish legal union, Craig appeared not only to be ignorant of English law but also to be waving a red flag in front of English MPs and lawyers, who remained as committed as ever to resisting any sort of 'reception' of the civil law. Craig, however, was not either unaware of or insensitive to English attitudes towards the civil law. He recognized that the civil law was not 'in vogue' in England and that English lawyers preferred merely to 'salute it from the threshold' rather than to use it as a basis of their decisions.[39] Nevertheless Craig was convinced, after reading Bracton and the decisions reported by Dyer and Plowden, that many points of English law had in fact been determined by the civil law. English judges, he wrote, just like their Scottish counterparts, 'determine the cause on arguments drawn from the *jus civile*', especially when the law was uncertain and differences of opinion arose.[40] In other words, there *had* been a reception of civil law in England, even if no one in England would admit it. And as far as Craig was concerned, a much more thorough reception could take place in the future, since all English legal controversies were 'easily capable of solution by the civil law'.[41]

As a plan for legal union Craig's proposal did not have much chance of success. The one way to guarantee the continued separation of English and Scottish law was to propose their union on the basis of the civil law. As a commentary on English law, however, Craig's argument is of considerable interest and not without merit. No one could argue that England had experienced a 'reception' of civil law similar to the reception that had taken place in France, Germany, Italy, or for that matter Scotland, in the fifteenth

---

[39] Ibid. 326.
[40] Ibid. 312.
[41] Ibid. 327. On Craig and the civil law see A. Watson, *The Making of the Civil Law* (Cambridge, Mass., 1981), 56–61.

and sixteenth centuries. The law that had been received in those countries was the law of the Bartolists, those commentators who had interpreted the civil law of Justinian in the light of customary feudal law.[42] When Craig referred to the civil law, he was not thinking of this body of substantive law. For him, as for many of his contemporaries, the civil law meant the basic principles of justice and equity that were embodied in the Justinianic code and had become the foundation of *jus gentium*, the law of nations.[43] These principles were considered to be rational, at least in the legal sense of that word, and were believed to be in accordance with the basic principles of all law, the law of nature.

If one accepts this definition of the civil law, then a number of statements from seventeenth-century English lawyers can be adduced to support Craig's contention. In 1628 Serjeant Ashley claimed that *jus gentium* 'ever serves for a supply in the defect of the common law, when ordinary proceedings cannot be had'.[44] Sir John Doddridge, who was learned in the civil as well as the common law, argued that when new matter was considered whereof no formal law was extant, 'we do as the Sorbonnists and the civilians, resort to the law of nature, which is the ground of all law, and then, drawing that which is more conformable for the commonwealth, do adjudge it for law'.[45] Doddridge also showed that the maxims of the common law, many of which were borrowed directly from the civil law, provided a very close approximation of the law of nature, while Bacon insisted that these maxims might be employed 'in new cases and such wherein there is no direct authority'.[46]

Even the process of law-making in the English parliament, the citadel of anti-Romanist sentiment, was not immune to either direct or indirect civilian influence. Arguments drawn

---

[42] See S. E. Thorne, 'English Law and the Renaissance', *La Storia del dirritto nel quadro delle scienze storiche* (Florence, 1966), 437–45.

[43] It was not uncommon for 16th-century Scottish lawyers to view the civil law in this way. See P. Stein, *Roman Law in Scotland*, Pt. 5, 13b of *Jus Romanum Medii Aevi* (Milan, 1968), 40. In *The Jus Feudale* (tr. J. A. Clyde, London, 1934), i. 28, Craig states specifically that 'in our own realm of Scotland we use the Roman law so far as it appears to us to be consonant with natural equity and reason'.

[44] *LJ* iii. 758.

[45] Harl. MS 5220, fo. 4ᵛ.

[46] J. Doddridge, *The English Lawyer* (London, 1631); P. Stein and P. Strand, *Legal Values in Western Society* (London, 1974), 101.

from the civil law were presented, for example, during the debates on the bill of naturalization in 1607—by all standards a doubtful case—and such discussion was opposed only when it appeared to touch the king's prerogative.[47] And later in the century, when parliament passed the Limitations Act, by which it provided a foundation for the doctrine of adverse possession, the drafters of the bill might very well have drawn upon the Roman doctrine of acquisitive prescription.[48] In other words, Craig might have been closer to the truth than he appears to have been when he claimed that the civil law had illuminated many English legal controversies.

Craig's hope that the civil law would provide a foundation for Anglo-Scottish legal union suggests one reason why so many English civil lawyers actively supported the cause for further union. Compared to the common lawyers, the civilians were a relatively small group who practised ecclesiastical law in the church courts and maritime law in the admiralty courts. Until the 1640s, however, when puritans, common lawyers, and parliamentarians almost succeeded in destroying their profession, they played a far more important role in public life than their small numbers would appear to warrant.[49] Their political prominence was most apparent when the Jacobean union was being negotiated and debated. Not only were two civil lawyers, Sir John Bennet and Sir Daniel Dun, selected as commissioners for the union in 1604, but both men partici-pated energetically in the parliamentary debates on the question of union in 1607.[50] One of their colleagues, Francis James, became the first MP to propose a 'perfect union' as an alternative to the plan of the commissioners,[51] while three other members of the profession—John Cowell, Alberico Gentili, and John Hayward—wrote treatises in support of the union.[52]

---

[47] *State Trials*, ii. 563; *CJ* i. 1020.

[48] See C. Donahue, Jr., 'The Civil Law in England', *Yale Law Journal*, 84 (1974), 181.

[49] See generally B. P. Levack, *The Civil Lawyers in England, 1603–1641: A Political Study* (Oxford, 1973).

[50] *Bowyer Diary*, 227–8, 305–6; *State Trials*, ii. 565.

[51] Notestein, *House of Commons*, 225.

[52] A fourth author, William Clerk, who did not take a degree in civil law but who was none the less learned in it, wrote in favour of the union. Trinity College Dublin, MS 635.

There are a variety of reasons for the civilians' prominence in the debates on the union. As men versed in international law, a branch of law that was closely allied to the civil law, the civilians were especially knowledgeable in matters concerning the relations between princes and states. For this reason they had been active in negotiating late sixteenth-century treaties between England and Scotland regarding the Borders,[53] and for the same reason their services were needed even more in 1604. Their legal education also conditioned them to think in terms of such matters as sovereignty, the union of states, and naturalization—the very issues that the union raised. Craig's book, however, suggests that there was another reason for the civilians' interest in, and support of, James's union project. In the early seventeenth century a number of English civilians, together with some common lawyers who had an interest in the civil law, set out to prove that the common law and the civil law, despite their substantial differences, were in fact compatible.[54] There was no better way to prove this compatibility than to show that Scots law, which had been infused with much Roman law, could be united with English law. Whether any of them wished to witness a partial English 'reception' of Roman law by means of Anglo-Scottish legal union is difficult to determine. The civilians never gave any indication of such a radical goal.[55] But there is no question that the idea of legal union based on the fundamental principles of the civil law appealed greatly to them.

The work of John Cowell, Regius Professor of Civil Law at Cambridge from 1598 to 1611, provides the best illustration of the strength of this appeal. Cowell was not as explicit as Craig in stating his hope for the eventual union of English and Scots law. He never actually referred to the possible creation of a single body of Anglo-Scottish law, perhaps because he was thinking more in terms of uniformity than of fusion. Nevertheless, Cowell wrote *Institutiones Juris Anglicani* (1605), in which

[53] W. Nicolson (ed.), *Leges Marchiarum, or Border-Laws* (London, 1705), 149.

[54] Levack, *Civil Lawyers*, 131–40; D. Coquillette, 'Legal Ideology and Incorporation I: The English Civilian Writers, 1523–1607', *Boston University Law Review*, 61 (1980), 1–89.

[55] D. Veall, *The Popular Movement for Law Reform 1640–1660*, (Oxford, 1970), 69, argues that Bacon wished for this type of reception. Bacon admired the civil law, but there is no evidence for a pro-reception position.

he placed the common law of England within the framework of the civil law, with the specific purpose of promoting Anglo-Scottish unity. Taking a much more practical approach to the problem than Craig, Cowell recognized that the first step to legal union would have to be the codification of both laws, that is, the systematic and rational arrangement of the laws under clearly defined headings. In the *Institutiones* he prepared such a code for England, while at the same time recommending the drafting of a similar code for Scotland. The latter task, Cowell admitted, would be no mean feat, since Scots law was an 'undigested mass'.[56] He need not have completely despaired, however, since Balfour's *Practicks* had at least made some progress towards the systematic arrangement of Scots law, and at the time Cowell wrote there was considerable sentiment in Scotland in favour of codification.[57] And the framework for codification that Cowell recommended, that of the civil law, could not be expected to arouse the same type of instinctive, negative reaction that it did in England, since the civil law already commanded great respect in the northern kingdom.

Cowell's work, therefore, was much more than an academic exercise. It was a serious attempt at the codification of English law on the basis of the civil law, and it represented a practical, albeit preliminary effort towards the realization of legal union. It directly inspired the Scottish unionist David Hume of Godscroft to propose the negotiation of an equitable legal union by a special ten-man commission,[58] and it experienced a revival at the time of the Cromwellian Union. In 1651 the English parliament, at the very time that it was making provisions for the administration of justice in Scotland, ordered a translation and new edition of the *Institutiones*. Although the new edition lacked its original preface, in which Cowell discussed Anglo-Scottish union, its reappearance can best be explained by its relevance to this question. The publication of

---

[56] Cowell, *Institutiones*, Epistola Dedicatoria.

[57] *The Practicks of Sir James Balfour of Pittendreich*, ed. P. G. B. McNeill (Stair Society, 1962–3); A. Williamson, *Scottish National Consciousness in the Age of James VI* (Edinburgh, 1979), 65–6. John Skene's compilation was completed after this date. *Original Letters Relating to the Ecclesiastical Affairs of Scotland*, ed. D. Laing (Bannatyne Club, 1851), i. 399.

[58] Hume, 'Tractatus Secundus', NLS, Advocates MS 31. 6. 12, fo. 52ᵛ.

this work by order of parliament did not lack irony, since Cowell had been the butt of parliamentary criticism for his absolutist views in 1607.[59]

Cowell's *Institutiones* was not the only plan for codification that appeared in connection with the plans for legal union. Sir Henry Hobart apparently drafted a brief summary of English law in order that it could be compared with a similar Scottish document in 1607, but his work has not survived.[60] Much better known is the work of Bacon, who even before the union had proposed the compilation of a 'Digest' of English law. In the early seventeenth century Bacon drafted a *Preparation Toward a Union of Laws*, in which he summarized the main elements of English public law.[61] Bacon's draft was much less ambitious than Cowell's, but in one sense it was more realistic, since it omitted all aspects of private law and it did not offend English sensibilities by using the civil law as its framework. Nevertheless, Bacon's work did not lead to any further efforts to bring the laws closer together. The only other discussion of the possibility of constructing a code of Anglo-Scottish law appeared at the beginning of the eighteenth century, by which time the chances of success had been greatly reduced. Seton, Fairfax, and Chamberlen all advocated the compilation of a digest of English and Scottish law, but none of these men actually did the type of preparatory work that Cowell and Bacon had undertaken a century earlier.

The various proposals for the codification of English and Scots law in preparation for legal union established a connection between the cause of law reform and that of the union. Those who wished to simplify, arrange, or change the law of either country in any significant way welcomed the union as a golden opportunity to implement their schemes. By the same token unionists often mentioned law reform as one of the attractions of further union. Thornborough, for example, mentioned the 'reformation of laws' as one of the benefits that would attend the perfection of the union. The union could be

---

[59] John Cowell, *The Institutes of the Lawes of England* (London, 1651). The Preface of the new edition refers to Cowell as a good 'commonwealth's man', sig. A2ʳ.

[60] HMC, *Salisbury MSS*, xix. 363.

[61] Bacon, 'Preparation' in *Works*, ii. 160–6.

expected to reduce the 'rigour' of English law while restraining the 'liberty' of the Scottish.[62] The author of *Rapta Tatio* argued that the Scots would be able to reform some abuses in their laws that had 'crept in by time, custom or misinterpretation', while the English would at the same time 'get the benefit' of Scottish laws.[63] King James, speaking to his English parliament in 1607, argued that the implementation of the union would allow parliament to 'amend the laws, polish them, and sweep off the rust of them'.[64] Even Sir Edwin Sandys, in advocating an imperialistic brand of 'perfect union', jumped on the bandwagon of law reform.[65]

III

The proposals made by law reformers, legal scholars, English civil lawyers, and statesmen from both nations to promote a union of English and Scots law encountered stiff opposition in both England and Scotland. The strength of this opposition was one of the main reasons why the Treaty of Union that was eventually approved in 1707 made specific provisions for guaranteeing the independence of the two legal systems. In arguing against legal union Englishmen and Scots often used the same language. Pamphleteers and public figures from both nations, for example, spoke of the threat to their 'fundamental laws'. The concerns of Englishmen and Scots in this regard were, however, somewhat different. Scots were mainly fearful of a loss of national sovereignty whereas Englishmen wished to prevent changes in a legal system that they considered to be second to none.

Most Scottish objections to legal union focused on the threat that such a union posed to Scottish sovereignty. This, as we have seen in Chapter 2, was also the main concern with the more general, constitutional provisions of the proposed union. The concern was not so much that Scots law itself would change; Scots law had always been open to outside

---

[62] Thornborough, *Discourse*, 18–19.

[63] *Rapta Tatio* (London, 1604), sig. GI. There is an echo of this argument in G. Mackenzie, Earl of Cromarty, *Parainesis Pacifica: or a Persuasive to the Union of Britain* (London, 1702), 6–7.

[64] *Political Works of James I*, 292–3.

[65] Notestein, *House of Commons*, 246.

influences. The real danger was that English law would be imposed on Scotland, much in the same way it had been imposed on Wales, and Scotland would therefore be deprived of its independence as a kingdom. The Scottish lawyer John Russell stated the case most clearly in the union treatise he composed for the benefit of King James in 1604. Russell strongly favoured the cause of union, but he was unalterably opposed to any attempt to change the fundamental laws of his kingdom. 'Shall a free kingdom,' wrote Russell, 'possessing so ancient liberties, become a slave?'[66] The same concern underlay the restrictions that the Scottish parliament imposed on the commissioners that it appointed to negotiate the union in 1604. The commissioners were instructed not to tolerate any derogation of the 'fundamental laws, ancient privileges, offices, rights, dignities and liberties' of their country.[67] The phrasing of this clause is significant in two respects. First, what mattered were the fundamental laws, the basic laws by which the kingdom operated, not the specific provisions or procedures of Scots law. Second, the commissioners were instructed to preserve not only these fundamental laws but also the 'privileges, offices, rights, dignities and liberties' that Scotland had—in other words, all the characteristics of a sovereign nation.[68]

Scottish fears of a loss of sovereignty were so great that even those Scots who favoured a union of laws expressed appropriate caution. As mentioned above, Craig, whose enthusiasm for a union of laws could not be contained, none the less recommended against its implementation, not only because he wished to support the commission of 1604 but also because he feared a loss of Scottish independence. David Hume also found the idea of legal union attractive but still recommended that the laws remain as they were for the time being.[69] Throughout his union treatises Hume insisted that the union should be equitable, and it was doubtless his fear of a one-

---

[66] J. Russell, 'Ane Treatise of the Happie and Blissed Unioun', in *Jacobean Union*, 89.   [67] *APS* iv. 264.

[68] In 1706 a Scottish lawyer claimed that 'our Church constitutions, our laws concerning private rights, our judicatories, etc, are our greatest rights, and peculiar privileges, which can neither be thrown away, nor taken from us, without the inevitable ruin both of the Church and Nation'. *The Trimmer*, 5.

[69] Hume, 'Tractatus Secundus', f. 52ᵛ.

sided legal union that led him to call for a temporary maintenance of the legal status quo.

Scottish fears concerning a loss of national sovereignty also
influenced the question whether English or Anglo-Scottish
courts would have any jurisdiction over Scottish cases. This
matter became an issue only in the late seventeenth and early
eighteenth centuries, when an incorporating union became
likely. In 1670 the Scottish commissioners for the union
demanded that all Scottish legal processes be handled in
Scotland, and in 1706, the great Scottish patriot Andrew
Fletcher of Saltoun expressed his fear that the 'judicatories for
administering justice and the cognizance of all law suits shall
be carried up to London, either by first instance or by way of
appeal'.[70] The fear that Scotland would actually lose its courts
had no foundation, but the possibility of appeals going to
London was real enough to lead George Mackenzie, the Earl
of Cromarty, who favoured an incorporating union, to
propose that appeals to the proposed British parliament be
prohibited on the grounds that most of its members would
be 'absolutely ignorant of, and strangers to our laws'.[71] Once
again, the real issue was one of sovereignty. According to
Cromarty, the Treaty of Union would leave Scotland
nothing of sovereignty . . . to reside amongst us but our
sovereign courts of judicature'. It was imperative therefore,
not only that appeals be banned but also that the reform of the
Court of Session, the highest Scottish civil court, be undertaken by a committee of Scottish MPs, sitting independently
of the whole parliament.[72] And although Cromarty was willing
to concede the power of regulating the Court of Session to the
British parliament, he argued that this should not be done
without consulting the Scottish judges to determine whether
the regulations amounted to alterations.[73]

---

[70] *Cromwellian Union*, 203; [Fletcher], *State of the Controversy*, 20.

[71] [Mackenzie], *A Letter to a Member of Parliament*, 5. See also [T. Spence], *The Testamentary Duty of the Parliament of Scotland with a view to the Treaty of Union* (1707), 9.

[72] Ibid. For a similar plan, according to which the Scottish Estates would meet to consider proposed changes in private law, the constitution of the church or the form of judicatories, see *The Trimmer*, 7–8. Authorship of this pamphlet has been attributed to John Spottiswoode or Robert Whytt of Bennochy, both of whom were advocates.

[73] [Mackenzie], *A Letter to a Member of Parliament*, 7–8. On the reform of the Court of Session see N. Phillipson, 'Nationalism and Ideology', in *Government and Nationalism in Scotland*, ed. J. M. Wolfe (Edinburgh, 1969).

Scottish arguments against legal and judicial union revolved, therefore, around the fundamental question of Scottish sovereignty. In England, however, where the union never threatened to destroy sovereignty as it did in Scotland, opposition to legal union centred on the undesirable changes that would thereby take place in English law. Many seventeenth-century Englishmen believed that the common law was not only the best system of law known to man but that it was also changeless. This opinion derived mainly from the common lawyers of Sir Edward Coke's day, who claimed that the common law was a body of immemorial custom, unchanged since ancient times.[74] This myth became a stumbling block for law reformers, and it forced legal unionists to try to debunk it. In his treatise on the union Hayward attacked as 'fabulous' the view that 'the laws of England, since the time of Brutus, were never changed'.[75] No wonder that Cecil, in giving instructions to parliament regarding the union, concluded, 'Caution where laws are to be changed.'[76]

If English lawyers were cautious of any legal change whatsoever, they were especially wary of the attempted introduction of any foreign legal influences. Unlike Scots law, which had incorporated large amounts of Roman and canon law, English law had developed along distinctly national lines. It had already become a mature legal system in the twelfth century, before Roman law had begun to influence the secular law of other European countries. England therefore had little need for Roman law and soon began to take pride in its independence of it. When changes modelled on Roman law were proposed in England, they encountered deliberate resistance. The most famous statement of English legal conservatism, '*Nolumus leges Angliae mutari*', which became a shibboleth of lawyers and MPs at the time of Coke, received first formal mention in parliament when the Lords considered the introduction of the Roman and canon law doctrine of legitimation *per subsequens matrimonium*. Even in the early seventeenth century, proposals for legal change were associated with the possible introduction of Roman influences, and they

---

[74] Pocock, *Ancient Constitution*, 30–55.
[75] Hayward, *Treatise of Union*, sig. A1ᵛ.
[76] HMC, *Salisbury MSS*, xix. 414.

were refuted on the grounds of the superiority, if not the perfection, of the existing common law. During the debate on the union in 1604, an anonymous MP, paraphrasing Fortescue, argued against the king's project in these words:

There cannot be a complete union (whatsoever is pretended) without change of laws, customs, etc. . . . Yet could Englishmen never be drawn to alter the laws and customs of this kingdom, which if they could have needed bettering by change, some of those kings [who had conquered England] . . . would have changed them or intermixt them with others of their own, especially the Romans, who judged all the rest of the world (England excepted) by their own laws and ordinances, only preserving the English customs entire and untoucht as more agreeable to this country and commonwealth than any other and more willingly obeyed than any other.

Sir Thomas Smith, as he was principal Secretary to two of the worthiest princes of this land (and therefore could not b[e] any way ignorant of our laws and customs and of the English Government), so he was also a Doctor of the Civil Laws, and therefore (by all likelihood), if he had found any defect in our laws or usages (fit to be amended or altered by any commixtion with his own professed law), he would have strived to have impaired the one as it stood alone and unmixt and have drawn some help fom the other (which was the civil law). But he (on the contrary part) in his treatise *De Republica Anglorum* by plain words testifieth that there was never in any commonwealth devised a more discreet, a more delightsome and gentle, nor a more certain way to rule the people, whereby they are kept always as it were in a bridle of good order, and sooner prevented and looked unto that they should not offend than punished when they have offended.[77]

The speaker went on the invoke Aristotle's stern warning against tampering with the ancient laws and thereby threatening the ruin of the entire state, and he concluded with a strong recommendation against any complete or perfect union, 'which must necessarily draw on at last alteration or mutual participation of laws'.[78]

The reference to the civil law in this anti-unionist speech had a special relevance to the issue at hand, since Englishmen

[77] Harl. MS 1314, f. 15ᵛ. For the passage from Fortescue see *De Laudibus Legum Anglie*, 39.
[78] Harl. MS 1314, f. 16ᵛ. For a similar view by an opponent of legal union a full century later see P. Paxton, *A Scheme of Union between England and Scotland* (London, 1705), 8.

frequently identified Scots law with the civil law.[79] The
identification naturally served as an argument against legal
union. If Scots law were essentially civil law, then legal union
was simply out of the question, for the union would raise the
possibility of a partial reception of the civil law in England.
The only problem with the argument was that Scots law was
not essentially civil law. Scots law had, it is true, absorbed
large amounts of both Roman and canon law in the fifteenth
and sixteenth centuries. It had even received the principle of
legitimation *per subsequens matrimonium* that England had so
emphatically rejected. The Scottish reception of civil law,
however, was hardly complete. Scots law in the early seventeenth century was, as it remains today, a mixed legal system,
one in which certain tenets of civil law had been grafted on to
an essentially native core of customary, case, and statutory law.[80]
    In trying to strengthen the case for legal union James I tried
to place the civil law component of Scots law in perspective.
Borrowing from Sir Henry Savile and other scholars, James
asserted that there were three main types of Scottish law: the
civil law, statute law, and the laws for tenures, wards, liveries,
seignories, and lands. The last, according to James, was
basically English law that King James I of Scotland had
imported in the fifteenth century. Credit for the introduction
of the civil law into Scotland also went to James I, who
'brought it out of France' by establishing the Court of Session
on the model of the Court of the Parliament of France'. The
judges of the Court of Session did not, however, 'govern
absolutely by the civil law', as in France. Whenever there was
a clear solution to a legal problem in Scots law, they were
obliged to rely upon it, having recourse to the civil law only
when the municipal laws were defective.[81] His conclusion was
the *neither* country was subject to the civil law and therefore
legal union was in fact possible, an argument that looked a lot
different from that of Craig, who based his hopes for a legal
union on the fact that *both* laws conformed to the civil law.

[79] *Somers Tracts*, ii. 135; Sir H. Spelman, 'Of the Union', in *Jacobean Union*, 180;
Beaulieu Palace Library, Papers on Scotch Affairs, III, item 1.
[80] On civil law in Scotland see Stein, 'Roman Law in Scotland', and A. Watson,
Legal Transplants: An Approach to Comparative Law (Edinburgh, 1974), 44–56.
[81] *Political Works of James I*, 301–2.

James might have strengthened his case if he had shown, as
Savile did, that even in France, where James claimed that the
civil law was absolute, the situation was little different than in
Scotland, since even in France the civil law was only used as a
guide to the interpretation and disputation of native custom.[82]

IV

Even if James had succeeded in dispelling English fears of a
reception of the civil law via Scotland, he had not dealt with
the most formidable obstacle of all to legal union—the
substantial differences that did in fact prevail between the two
legal systems. Like so many of his fellow unionists, James
assumed that since Scotland had originally borrowed its
common law from England (he had actually gone so far as to
claim that the Scots had 'none' of their own except what they
took from England), there could be little difficulty in reconcil-
ing the two. Perhaps he can be forgiven his preoccupation
with the origins of Scots law, since contemporary legal theory
in England placed much greater emphasis on the origins than
on the development of the law. But like it or not, both Scots
law and English law had developed significantly since the
Scottish reception of English law, and their development had
not always occurred along parallel lines. Craig had a
somewhat better understanding of the differences that had
arisen, but even he was not fully aware of their extent.[83] Later
in the seventeenth century, however, Sir Matthew Hale, who
like James and Craig recognized the original identity of the
two laws, pointed out how widely the two laws had diverged
since the kingdoms had become separated. In his *History of the
Common Law of England* he wrote: 'We have but few of their
laws that correspond with ours of a later date than Edward I
or at least Edward II.'[84]

There was much truth in Hale's statement. By the time of

[82] Savile, 'Historicall Collections', in *Jacobean Union*, 233–5. Roman law did not
possess positive authority in any European state. See K. Luig, 'The Institutes of
National Law in the Seventeenth and Eighteenth Centuries', *JR*, NS 17 (1972),
200–1.

[83] W. S. McKenzie, review of *De Unione Regnorum Britanniae* in *SHR* 7 (1910), 301.

[84] Sir M. Hale, *The History of the Common Law of England*, ed. C. M. Gray (Chicago,
1971), 132.

the Union of the Crowns, substantial differences had arisen between English and Scots law, especially in the law of persons and of property. Some of these differences can be attributed to the Scottish reception of Roman and canon law, but others can be traced to legal developments in England rather than in Scotland. This is especially true in the area of land law, where English statutes, enforced by legal decisions, had made dramatic changes that found no parallel in Scotland. It should be noted that even as late as the early seventeenth century, acts of parliament did not play the same role in the development of the law in Scotland as in England. Scottish statutes rarely if ever brought about substantial changes in the law; they usually simply confirmed developments that had already taken place in the courts. Consequently Scots law, which paradoxically had never assigned as much importance to precedent as had English law, had changed more gradually. There is more than a little irony in the fact that Scots law, which has remained open to outside influences such as the civil law, and which has never been the subject of a myth of changelessness and immemoriality, had in fact changed less than the allegedly immutable and strictly national English common law. Even during the period of English hegemony in Scotland, when English law exercised a profound influence over Scots law, Scottish parliaments did not make changes in the land law corresponding to those brought about by the English statutes *De Donis Conditionalibus* (1285) and *Quia Emptores* (1290), which in effect ended the practice of subinfeudation in England. And after the two countries had become completely independent of each other, the English law of entails followed a completely separate line of development from that of Scotland, receiving its most significant changes through the Statutes of Fines (1484 and 1489) and the Statutes of Uses (1536) and Wills (1540), enactments that had no parallel in Scotland.[85]

Just as the substantive laws of the two kingdoms had grown so far apart by the seventeenth century that the subjects of

---

[85] R. Burgess, *Perpetuities in Scots Law* (Stair Society, 1979), 24–73; A. W. B. Simpson, 'Entails and Perpetuities', *JR*, NS 24 (1979), 1–20; C. A. Povolich, 'Scottish Juridical Institutions and Practices under the Impact of the Union with England', Ph.D. thesis (Univ. Southern California, 1953), 158–9.

each could freely admit that they were ignorant of the law of the other,[86] so the procedural laws had also become distinct. In civil proceedings the main difference was the absence of a Scottish trial jury. Scottish courts had never recognized the utility or necessity of this means of settling civil disputes. In fact, some Scots objected strenuously to the English practice of entrusting an illiterate jury with the 'pernicious power' of determining the 'important points of right and wrong'.[87] Scottish subjects, therefore, did not have the benefit of what Englishmen considered to be one of their most cherished rights. Englishmen were so jealous of their right to civil jury trials that an early seventeenth-century proposal to create courts of conscience in London to determine suits involving claims of less than £10 met with strong opposition on the grounds that it would take away 'the good, ancient, approved law of England, to wit, trials by jury and introduce an arbitrary jurisdiction'.[88] Eventually Scottish practice was brought into line with the English in this regard. Scottish Whig agitation for the civil jury trial, coupled with the resentment of the Lords in having to determine the facts of the cases brought to them on appeal, resulted in the introduction of the civil jury in Scotland in 1815.[89] The difficulty in securing the passage of this legislation more than a century after the Act of Union had made it technically possible provides a good indication of the obstacles that stood in the path of legal union.

Differences between English and Scottish criminal procedure also constituted an impediment to legal union, especially since a union of laws would have begun with a unification of at least some of the criminal law of the two countries. Superficially the two systems of criminal procedure were not all that different. In both countries, for example, the determination of the facts of a case was left to a petty jury of laymen.

---

[86] *The Earl of Stirling's Original Register of Royal Letters Relative to the Affairs of Scotland and Nova Scotia from 1615 to 1635*, ed. C. Rogers (Edinburgh, 1855), i. 5; *Notes of the Treaty Carried on at Ripon between King Charles I and the Covenanters of Scotland, A.D. 1640* (Camden Society, 1869), 6.

[87] Fairfax, *Discourse*, 57.

[88] SP 15/36/27.

[89] 55 George III, c. 42, N. Phillipson, 'Scottish Public Opinion and the Union in the Age of the Association', in *Scotland in the Age of Improvement*, ed. N. T. Phillipson and R. Mitchison (Edinburgh, 1970), 136–7.

Nevertheless, Scotland did not have a presenting jury, which meant that it was easier for Scottish legal officials to initiate prosecutions than it was for their English counterparts. Scotland also had a public prosecutor in the person of the Lord Advocate, and it placed much greater emphasis on the value of written depositions taken during pre-trial examinations.[90] In the late seventeenth and early eighteenth centuries, moreover, Scottish trial juries (which unlike English juries could convict by majority vote) surrendered many of their powers to the judges and for a while did not even return general verdicts of guilty or not guilty.[91] All these procedural differences gave the impression that Scottish criminal law was more severe and arbitrary than that of England, and the fact that Scottish prisoners could not peremptorily challenge jurors or plead benefit of clergy served only to strengthen that impression. The only advantages that Scottish prisoners had over their English counterparts were the right to legal representation and the right to challenge the judge.[92]

Above and beyond the differences in both the substantive and procedural laws of the two countries, there were basic differences in English and Scottish attitudes towards the sources of the law and the way in which the law should be determined in particular cases. English parliaments may have been bolder in their use of statute to make new law, but Scottish courts had much greater freedom in determining what the law was. Thus English law, which in some respects was more creative than Scots law, appeared to be more rigid, mainly because it was much more tightly bound by precedent. Scottish judges had much more freedom than their English counterparts. In reaching decisions they could cite old acts of parliament, civil law, the auld Scottish laws, and the works of the Scottish institutional writers as well as precedent. They could also ignore unrepealed statutes that were considered to

[90] *Selected Justiciary Cases*, ed. J. I. Smith (Stair Society, 1972–4), ii. p. x.

[91] I. D. Willcock, *The Origins and Development of the Jury in Scotland* (Stair Society, 1966), 218–21.

[92] On the differences between English and Scottish criminal procedures see B. P. Levack, 'English Law, Scots Law and the Union, 1603–1707', in *Law-Making and Law-Makers in British History*, ed. A. Harding (London, 1980), 113–14; Sir G. Mackenzie, *A Vindication of the Government of Scotland during the Reign of King Charles II* (Edinburgh, 1691), 18.

have fallen into 'desuetude', a legislative status unknown to England. Judges in Scotland had so much latitude that they in fact determined the nature of Scottish 'custom'. Custom in Scotland meant not only immemorial local or general custom, as it did in England, but 'learned custom', a body of law that had developed through judicial decisions and practice.[93]

The greater freedom and flexibility of Scots law were also apparent in its laxity regarding the use of proper writs and forms. In England a plaintiff could not receive judgment unless he could determine the proper writ and then adhere to the proper form in filling out his complaint. In Scotland, on the other hand, the right to judgment took precedence over the existence and use of the proper remedy, and therefore pleadings were not restricted to stereotypical formulas. Judgments in Scottish courts, unlike those in England, were based more on the substance than the form of the dispute.[94]

Yet another example of the greater flexibility of Scottish justice was the fact that Scottish courts could administer both law and equity, whereas in England separate courts were entrusted with the two types of jurisdiction. Whether the Scottish arrangement in this regard was superior to the English is open to dispute, but it should be noted that the Scottish system became the model for various programmes of law reform in England and colonial America and was eventually implemented in England when the High Court was established in 1873.[95]

Substantive and procedural differences in the two laws, combined with their different sources and the different principles upon which judgments were made, led naturally to two different systems of legal education. No greater distinction between the legal world of London and that of Edinburgh in the seventeenth century can be found than in the education, training, and general legal outlook of the two legal professions. In keeping with the cosmopolitan character of Scots law, its

[93] On Scottish custom see T. B. Smith, 'Authors and Authority', *Journal of the Society of Public Teachers of Law*, NS 12 (1972), 9; J. T. Cameron, 'Custom as a Source of Law in Scotland', *Modern Law Review*, 27 (1964), 306, 318.

[94] A. G. Murray, 1st Viscount Dunedin, *The Divergencies and Convergencies of English and Scottish Law* (Glasgow, 1935) 20–7; Povolich, 'Scottish Juridical Institutions', 5.

[95] B. Bailyn, *The Ordeal of Thomas Hutchinson* (Cambridge, Mass., 1974), 96; 36 Vict., c. 66.

reception of the civil law, and its greater respect for right than for writ, Scottish lawyers generally received their legal education on the Continent, where they studied the civil law.[96] Their legal education, therefore, was largely jurisprudential. There was no provision for the formal study of Scots law until 1722, when a chair in that field was established at the University of Edinburgh.[97] Prior to that time knowledge of Scots law was usually gained through informal means, such as by apprenticeship or association with senior lawyers in the Faculty of Advocates.[98] Even then, however, the Faculty of Advocates did not require a knowledge of Scots law for admission as an advocate until 1692. In England, on the other hand, formal legal education consisted almost entirely in professional training in the common law, which took place at the Inns of Court in London. An English lawyer might have gained some knowledge of jurisprudence or the civil law by taking an earlier degree in civil law at Oxford or Cambridge or through private reading, but the legal education he needed for admission to the bar was both municipal and practical.

In a few minor respects the English and Scottish legal professions began to resemble each other more closely during the seventeenth century. By an order of James VI Scottish lawyers donned the black gowns of their southern counterparts, while later in the century the Faculty of Advocates acquired much of the cohesion and control over its membership that the Inns of Court already possessed.[99] Nevertheless, these superficial similarities could not disguise the fact that the members of the two legal professions, in addition to practising different bodies of municipal law, also represented two strikingly different legal traditions. As long as English and Scottish lawyers continued to receive different types of legal training and education, legal union would be difficult, if not

---

[96] Watson, *Legal Transplants*, 46; R. Feenstra and C. J. D. Waal, *Seventeenth-Century Leyden Law Professors and their Influence on the Development of the Civil Law* (Amsterdam and Oxford, 1975), 84–5; N. T. Phillipson, 'Lawyers, Landowners, and the Civic Leadership of Post-Union Scotland', *JR* NS 21 (1974), 107; [Mudie], *Scotiae Indiculum*, 16.

[97] *An Introduction to Scottish Legal History* (Stair Society, 1958), 61.

[98] N. Wilson, 'The Scottish Bar: The Evolution of the Faculty of Advocates in its Historical Social Setting', *Louisiana Law Review*, 28 (1968), 237, 240.

[99] *RPCS* viii. 612–14; Wilson, 'Scottish Bar', 237.

impossible to achieve. Indeed, it has been suggested that even the moderate anglicization of Scots law that took place in the nineteenth century could not have succeeded until those Scottish lawyers who continued to acquire their legal education abroad began to study law at Oxford and Cambridge rather than on the Continent.[100]

V

The strength of English and Scottish objections to legal union, together with the practical difficulties that any efforts to unite such different legal systems would have encountered, ensured that when the two kingdoms were incorporated into one polity in 1707, most of their laws remained distinct. The chances of such a union were never really very strong, and they became weaker as the seventeenth century progressed. The only time that a negotiated union of laws had even a remote chance of success was in the early seventeenth century. This was the time when most plans for legal union were presented and when some preparatory work was done. It was also the time in which Scots law bore a close enough resemblance to English law that lawyers, judges, and members of parliament could reasonably discuss the possibility of reconciliation. After that time, when Scots law entered the age of Stair and Mackenzie, it became more distinct from English law and acquired its own national character. Not only did these institutional writers give Scots law a coherence that it had not previously possessed, thus making it more resistant to potential English influence, but they specifically rejected parts of old Scots law that reflected previous English influence, such as the *Regiam Majestatem* of the thirteenth century.[101] For the first time in its history Scots law appeared to be completely independent of English law.[102]

This major development within Scots law made the practical difficulties involved in achieving legal union much more formidable and virtually ensured that a thorough legal union

---

[100] Ibid. 238.　　　　　　　　　　　　　　[101] *Practicks of Balfour*, i. xxxix.
[102] English law did, nevertheless, serve as one of Stair's sources. See W. D. H. Sellar, 'English Law as a Source', in *Stair Tercentenary Studies* ed. Walker, 140–50.

would not take place. Indeed, the only way that such a union could have been implemented, short of the creation of a new bar and bench trained in a new composite law,[103] was by the introduction of English judges into Scotland. Cromwell had used this tactic in the mid-seventeenth century, but it had required the use of force and even then had not brought about extensive legal change or anything approaching real legal union. To have undertaken such a programme in the early eighteenth century would have been unwise, unnecessary, and probably impossible. Povolich has argued that any attempt to amalgamate the two systems would have produced chaos in both nations, while the replacement of one system with the other would have uprooted basic judicial principles and wrought complete havoc in litigation.[104]

The Treaty of Union of 1707 guaranteed that a union of English law and Scots law would not take place. It did not, however, fully protect the integrity of Scots law or the autonomy of the Scottish judicial system. The Treaty preserved Scottish private law, but it did not declare it to be inviolable. Quite to the contrary, the Treaty allowed Scottish private law to be altered 'for evident utility of the subjects within Scotland'.[105] The Treaty preserved Scotland's separate judicatures, but it did nothing to prevent appeals from going to the English-dominated House of Lords. The Scottish commissioners did not take a stand on this issue, either because they assumed appeals would be infrequent or because they were reluctant to establish a new Scottish court of appeal. In any event, the Treaty was silent on this issue, since the English Lords did not wish to reopen a controversy with the House of Commons regarding the Lords' right to hear appeals of equity cases.[106] The Lords did not wait long, however, to establish their appellate authority in civil cases once the Treaty had been ratified.

[103] See the proposal of Chamberlen, *Great Advantages*, 20. Lord Belhaven feared that after the union Scottish judges would be 'laying aside their practicks and decisions, studying the common law of England'. J. Hamilton, 2nd Lord Belhaven, *The Lord Belhaven's Speech in Parliament the second day of November 1706* (1706), 1–2.

[104] Povolich, 'Scottish Juridical Institutions', 3.

[105] Pryde, *Treaty of Union*, 95.

[106] A. J. MacLean, 'The 1707 Union: Scots Law and the House of Lords', in *New Perspectives in Scottish Legal History*, ed. Kiralfy and MacQueen, 50–75.

In the area of public law the Treaty allowed as much legal union as the interests of the state would require. It is true that most of the criminal law of Scotland remained distinct from that of England, and with the exception of the law of treason and new crimes, which were defined by statute, it remained unaffected by English law, mainly because the Lords did not hear criminal cases on appeal after 1713.[107] But those aspects of public law which were constitutional in nature, not to mention those which affected trade and the customs, were drastically altered. The net effect of the Treaty was to establish a legal pluralism, but it was a limited pluralism which denied Scotland complete legal independence.[108]

Since the Treaty of 1707 did not fully preserve the sovereignty of Scots law or the autonomy of Scottish judicatories but instead subjected them to statutory and appellate control, it is easy for one to see the concessions of the Treaty to Scottish legal separatism as sops to Scottish national sensibilities. Indeed, one can argue that the Treaty deprived Scotland of all effective sovereignty and allowed England to control the country without having to assume the obligation of supervising the Scottish judicial system. Whenever English interests appeared to be threatened, Scots law could be changed accordingly. The complete reform of the Scottish law of treason in 1708, a process that resulted in the virtual imposition of English law on Scotland, proved that the English did in fact have the upper hand.[109]

Even if the English did have the power to change Scots law, the preservation of a large measure of Scottish legal independence within the union was of great benefit to Scotland. Separate courts and a separate legal profession served as focal points of Scottish national identity and provided a framework for independent political activity. Not only did the law courts physically occupy the space vacated by the defunct Scottish parliament, but lawyers replaced peers as the most prominent

[107] A. L. Murray, 'Administration and Law', in *The Union of 1707: Its Impact on Scotland*, ed. T. I. Rae (Glasgow, 1974), 46.

[108] On English influence on Scots law after the union see T. B. Smith, *Studies Critical and Comparative* (Edinburgh, 1962), 97–136; A. D. Gibb, *Law from over the Border: A Short Account of a Strange Jurisdiction* (Edinburgh, 1950), 1–2.

[109] 7 Anne, c. 21. See BL, Stowe MS 158, f. 91ᵛ for legal problems created by this statute.

figures in Scottish political life. The families that dominated
Scottish life in the eighteenth and nineteenth centuries were
the great legal families, while the most important officeholder
in the country was the Lord Advocate. Unless the Treaty of
Union had preserved an independent Scottish judiciary,
lawyers could not have served this important political and
national function. The only other group that might have done
so was the clergy, but in an increasingly secular age it was
unlikely that they could have exercised so much power and
influence. In a very real sense, the law courts and the legal
profession were all that the Scottish nation had.[110]

The preservation of Scots law and an independent Scottish
judiciary within the union possessed great significance for
England as well as for Scotland. By the Treaty of Union the
English common law accepted its first clearly defined, territorial limits within the king's dominions. As the English state
had expanded in the past, it had extended the common law to
those areas that it had incorporated. By the end of the
sixteenth century this process of legal incorporation had been
completed. The English state had also exported the common
law to those territories that it had settled overseas, such as
Ireland and the American colonies. When, however, it
became time to incorporate Scotland into a broader British,
but still predominantly English polity, the common law was
not extended to the newly incorporated area. The problem
was that in this case the common law encountered a legal
system that was already well established and, by 1707, highly
developed. The common law may have been easily export-
able, but it did not interact well with other legal systems. Just
as it was later to do in Canada and South Africa, therefore, the
common law chose to stay aloof from the law with which it
had come into contact.[111]

The recognition of a legal pluralism within Britain in 1707
gave the British state of the eighteenth century a peculiar
character. On the other hand it had one of the most important

---

[110] On lawyers, the law and Scottish nationalism see T. B. Smith, 'Scottish
Nationalism, Law and Self-Government', 34–51; A. Murdoch, 'The Advocates, the
Law and the Nation in Early Modern Scotland', in *Lawyers in Early Modern Europe and
America*, ed. W. Prest (London, 1981), 147–63.

[111] Sir L. Scarman, *English Law: The New Dimension* (London, 1974), 9–10.

characteristics of a unitary state. It possessed a parliament that passed legislation for all Britain and did not share its power with any regional or national assemblies. On the other hand the British state lacked the legal and judicial cohesion of the unitary state. Unlike both the English and Scottish states of the seventeenth century, the British state had no centralized judicial system and no single body of national law. At the same time the British state lacked another feature of both the English and Scottish states of the seventeenth century: a single state church. It is to this second British modification of the concept of the unitary state that we now turn.

4

Religious and Ecclesiastical Union

THE ecclesiastical terms of the Union of 1707, embodied in the two Acts of Security that were appended to the Treaty, in many ways paralleled its legal provisions.[1] Just as the Treaty created a British state in which two separate and independent legal systems coexisted, the two Acts of Security ensured that the new state would encompass two separate and independent national churches. This ecclesiastical pluralism, which was something quite different from religious toleration, was more complete, unconditional, and unequivocal than the legal pluralism that the Treaty recognized. For whereas the British parliament reserved the right to alter Scots law when 'evident utility' for Scottish subjects could be demonstrated, it possessed no corresponding power to encroach upon the integrity and established constitution of the Scottish church.

The two Acts of Security were appended to the Treaty in order to guarantee that the silence of the Treaty regarding ecclesiastical matters would not be misconstrued as a licence to bring the two churches into closer conformity. The Acts, therefore, constituted a final rejection of the various plans for ecclesiastical union that both Englishmen and Scots had proposed since the Union of the Crowns. These plans had attracted much more interest and support before 1660 than after, but they still remained the subject of considerable

[1] The Scottish Act of Security (Act for securing the Protestant Religion and Presbyterian Church Government in Scotland) 1706, c. 6., had to be inserted into any legislation regarding the enactment of a treaty of union with England. It became part of the Union with England Act of 1707 (c. 7) and it also was included in section 2 of the Union with Scotland Act which the English parliament passed in 1707 (6 Anne, c. 11). The English Act of Security (Act for Securing the Church of England as by Law Established) was also passed in 1706 (5 Anne, c. 5). The text of the two acts appears in Pryde, *Treaty of Union*, 105–12.

controversy in the early eighteenth century.[2] This chapter will explore the different proposals that were made for ecclesiastical union during the period 1603–1707. It will discuss the problems that such proposals entailed and the opposition that they attracted. Finally it will study the fate of the various attempts that were actually made to implement some sort of church union during the seventeenth century.

I

The phrase 'religious union', which frequently appears in seventeenth and early eighteenth-century writing about Anglo-Scottish relations, meant different things to different people. Sometimes it meant doctrinal unity, the profession by both nations of the same basic faith. When the phrase was used in this sense it often signified a unity that had to some extent already been achieved. Both England and Scotland during this period were Protestant nations, and although the Scottish church was more reformed or Calvinistic than the English, the differences between their official doctrinal creeds were not profound.[3] Bacon considered their union in this regard to be 'perfect' in 1604, while more than one hundred years later Daniel Defoe remained essentially of the same opinion.[4]

A second connotation of the phrase 'religious union' was a unity of ceremonies, the various liturgies employed in the religious services of the two nations. Achieving union in this regard would have involved the establishment of some sort of conformity between the liturgical practices of the two countries. The Scottish minister Alexander Henderson was thinking of this type of union when he expressed the hope that a Scot attending a service in England would feel as if he were

---

[2] The last of these was the proposal of William Seton for one body of ecclesiastical law in 1702. See *The Interest of Scotland*, 48–9.

[3] G. D. Henderson, *Religious Life in Seventeenth-Century Scotland* (Cambridge, 1937), 80–1.

[4] *L&L* iii. 223; [D. Defoe], *An Essay at Removing National Prejudices against a Union with Scotland*, Pt. I (London, 1706), 17. See also I. Bates, *Two (United) are Better than One Alone* (London, 1707), 17–18.

worshipping in his native Scottish parish.[5] A religious union of this sort was certainly within the realm of possibility, but the liturgical differences between the English and Scottish churches, which had been minimal in the 1560s and 1570s, had begun to widen by the beginning of the seventeenth century and continued to do so as the century progressed.[6]

A third type of religious union was ecclesiastical union or a union of church government. Since England and Scotland were least similar in this regard, proposals for this kind of religious union engendered the greatest amount of controversy. Most proposals for church union envisaged the creation of parallel, uniform structures of church government in both kingdoms,[7] but some actually anticipated their assimilation into one supranational ecclesiastical organization, in which there would be a single primate of Great Britain, or a single British convocation, synod, or general assembly. Another idea of church union that arose occasionally between 1603 and 1707 was the idea of a loosely structured, comprehensive British church under the monarch or parliament, in which the ecclesiastical institutions of each country would retain a great degree of autonomy.[8]

Just as the term 'religious union' had many different meanings, so the individuals who pursued this vague and elusive goal did so for a variety of reasons. In the most general sense it is possible to distinguish between those who considered religious union as part of a broader programme of Anglo-Scottish union and those who considered it as a goal in and of itself. Within the first group were a large number of unionists who recognized that there was little hope of achieving any sort of lasting or perfect union unless Englishmen and Scots considered themselves members of the same community or nation. In order to foster this sense of community, geographical proximity and an identity of languages were not

---

[5] J. D. Ogilvie, 'Church Union in 1641', *Records of the Scottish Church History Society*, 1 (1926), 155.

[6] G. Donaldson, *The Making of the Scottish Prayer Book of 1637* (Edinburgh, 1954), 7–29.

[7] See, e.g., BL, Stowe MS 187, f. 43, which refers to 'one form of ecclesiastical government', and M. R. Watts, *The Dissenters* (Oxford, 1978), 93.

[8] See, e.g., [J. Humfrey] *A Draught for a National Church Accommodation* (London, 1705); [Chamberlen], *Great Advantages*, 30–1.

sufficient; there would also have to a 'union of love', which membership in the same Christian community would be best able to provide. As John Doddridge wrote in 1604, 'Where there is no unity of religion, there can be no hearty love' and hence no perfect union.[9]

Many advocates of an incorporating or political union, the union of England and Scotland into one state, also saw religious union, especially a union in church government, as indispensable to the successful implementation of their designs. As long as contemporaries recognized and accepted the fact that the functions of church and state were inextricably intertwined, plans for a union of the English and Scottish states would entail plans for the union of the two churches. This attitude towards the union was always stronger in England than in Scotland, since the relationship between church and state was much closer in the southern kingdom. In England the state had established a degree of control over the church that had no parallel in Scotland, where the kirk had achieved a considerable amount of independence from the state and in the views of some was superior to it. Even in Scotland, however, there had always been close co-operation between church and state and persistent efforts by both ecclesiastical and civil authorities to influence each other.[10] Plans for political union in both countries, therefore, led naturally to the consideration of plans for church union. To those men who were accustomed to thinking in terms of a state church, the idea of two 'state churches' in one state seemed as unnatural as the prospect of two bodies with one head or two wives with one husband. Fletcher of Saltoun was absolutely correct when he observed critically in 1706 that there were many men who thought that 'the government of Church and State are so naturally interwoven that no nation can be at peace unless both these go hand in hand in their natural duties to each other and in their common dispensation to the

---

[9] Doddridge, 'Breif Consideracion', in *Jacobean Union*, 149. See also *Information from the Scottish Nation* (1640) in *Notes of the Treaty Carried on at Ripon*, 75.
[10] On church–state relations in Scotland see F. Lyall, *Of Presbyters and Kings: Church and State in the Law of Scotland* (Aberdeen, 1980), 1–22; J. Kirk, ' "The Politics of the Best Reformed Kirks": Scottish Achievements and English Aspirations in Church Government after the Reformation', *SHR* 59 (1980), 22–53.

whole members of the united society'.[11] To such men the ecclesiastical solution agreed upon in 1707 appeared to be a dangerous novelty.[12] But it was the very novelty of this solution that also led Dicey and Rait to regard the ecclesiastical provisions of the Treaty as a highly original piece of statesmanship.[13]

The person who most clearly associated a programme for ecclesiastical union with a broader plan for 'perfect' union was the main beneficiary of the Union of the Crowns, James VI and I. When he first outlined his union project, James indicated that it would include a union of the laws, churches, parliaments, and peoples of both kingdoms. As the various aspects of his programme met with resistance and defeat, however, James concentrated his efforts on achieving ecclesiastical union. His plan was to strengthen episcopacy in Scotland to such an extent that the two systems of church government would at least be uniform. In this undertaking his objectives were more political than religious.[14] The revival of Scottish episcopacy promised to strengthen his political control of Scotland, since the bishops were directly subordinate to him, and to allow for a closer co-ordination of English and Scottish policy. For James, therefore, religious union remained part of a broader union plan. It would provide a certain measure of *de facto* administrative union and still, it was hoped, lay the basis of the perfect union that James still desired.

The second large cluster of religious unionists consisted mainly of Scottish clerics who viewed religious union not as a concomitant or pre-condition of some broader union programme but as a goal in and of itself. Although English unionists occasionally supported the objectives of these men,

[11] [Fletcher], *State of the Controversy*, 11.

[12] HMC, *Mar and Kellie MSS*, 274; Robert Wyllie considered the prospect of two established churches in the same state as 'unnatural', since it would inevitably lead to the 'toleration' of one by the other. *A Letter concerning the Union*, 6.

[13] Dicey and Rait, *Thoughts on the Union*, 238–44.

[14] S. A. Burrell, 'The Covenant Idea as a Revolutionary Symbol: Scotland, 1596–1637', *Church History*, 27 (1958), 344. This is not to deny that James demonstrated a genuine commitment to Protestant and ultimately Christian unity. See W. B. Patterson, 'James I and the Huguenot Synod of Tonneins of 1614', *Harvard Theological Review*, 65 (1972), 241–70. In dealing with Scotland, however, his ecumenical goals appear to have been subordinate to his political objectives.

they did not share their exclusively or predominantly religious perspective, and they were usually unable to separate their plans for religious union from the more comprehensive schemes that Englishmen tended to favour. The most vocal of the Scottish religious unionists were a group of episcopalians who saw the Union of the Crowns as an opportunity for the new Constantine, a role now played by James VI and I, to lead the two nations, and eventually the entire world in a Protestant crusade to destroy the anti-Christ and establish the true religion. While a few moderate English puritans like Andrew Willet and Miles Mosse could also speak in terms of the two nations 'going up to Jerusalem together' under the direction of the new king,[15] many Englishmen had difficulty accepting this peculiarly 'British' version of apocalyptic thought, mainly because they had already subscribed to an interpretation of the apocalypse in which England alone, not a broader Britain, was the elect nation. Only in Scotland, where Arthur Williamson has shown that a corresponding version of Scotland as the elect nation did not and could not develop, was the vision distinctly 'imperial' or British.[16] John Knox provided the foundation of this Scottish thought, and it reached its full expression at the time of the Union of the Crowns in the work of such men as William Cowper, Patrick Galloway, and John Gordon. It also influenced the union writing of a Scottish layman, John Russell.[17]

The most interesting feature of this Scottish apocalyptic thought, aside from its distinctly episcopalian emphasis, is that it does not involve a specific commitment to further political, legal or economic union. What all these authors anticipate is the working out of a scriptural drama, the political prerequisites of which had already been realized in the Union of the Crowns. It mattered little whether the parliaments or privy councils of the two countries were assimilated, since the work of reformation was primarily the assignment of the new Constantine. In the case of Russell, the only Scot in this school who dealt directly with the non-religious aspects of union, there was in fact a coolness towards plans for

---

[15] A. Willet, *Ecclesia Triumphans* (Cambridge, 1603); M. Mosse, *Scotlands Welcome* (London, 1603).　　　[16] Williamson, *Scottish National Consciousness*, 1–47.
[17] Russell, 'Treatise of the Happie and Blissed Unioun', in *Jacobean Union*, 75–141.

political and legal union. The only Scotsman who subscribed to this type of apocalyptic thought while at the same time pursuing union in the broader sense was James himself, who obviously favoured projects that assigned him such a dramatic role in human history but who at the same time wished to unite his two kingdoms in the most complete manner possible.

The other Scots who viewed union in distinctly religious terms were presbyterians eager for world-wide reformation. To label these men unionists requires considerable qualification and explanation, since Scottish presbyterians often distinguished themselves throughout this entire period as die-hard opponents of both religious and political union. Indeed, Scottish presbyterians of the late sixteenth century, such as John Napier and Andrew Melville, had most emphatically rejected the unionism of Knox and his adherents. Denying both the sacred role of the first Constantine and the instrumentality of the civil government in achieving further reformation, these men rejected the claims made in favour of James as the new Constantine and the unionism that such claims implied.[18] The same presbyterians became the most determined opponents of the union project of James I, while their early eighteenth-century successors accounted for a large part of the anti-unionist agitation that arose in Scotland in 1706 and 1707.[19] At both times the presbyterians took an anti-unionist stance because they feared that the union would destroy the purity of their church. The presbyterians were not, however, opposed to union under all circumstances. In fact, at many different times throughout this period Scottish presbyterians argued strongly for a particular type of religious union, and in the eighteenth century they blamed the high church party in England for having sabotaged their efforts.[20]

The Scottish presbyterian programme for union originated in a version of the apocalypse according to which all nations, or at least the ten nations represented by the ten horns of the

---

[18] Williamson, *Scottish National Consciousness*, 86–96.

[19] Hetherington, *History of the Church of Scotland* (New York, 1859), 118; Lee, *Government by Pen*, 31–2; Pryde, *Treaty of Union*, 27–9; W. L. Mathieson, 'The Church and the Union', in *The Union of 1707: A Survey of Events*, ed. P. H. Brown (Glasgow, 1907), 55–9. Presbyterians also took the lead in opposing the proposed Cromwellian Union of 1651. Dow, *Cromwellian Scotland*, 39.

[20] [Ridpath], *Discourse*, 43–4; Daiches, *Scotland and the Union*, 72–3.

beast in Revelation, would achieve godly reformation.[21] The
result would be religious union, or at least religious uniformity, among the world's churches, and in particular between
the English and Scottish churches. The model of this union
was to be the fully reformed Scottish church. The Scottish
presbyterians were, therefore, articulating what we may call
the Scottish imperial model of British religious union. Instead
of accepting the imposition of English institutions, not to
mention English political and legal institutions, the Scots
hoped to export their own institutions to England, or at least
to encourage the remodelling of English institutions along
Scottish lines.

The earliest seventeenth-century expression of this type of
Scottish presbyterian unionism appears in the work of David
Hume of Godscroft.[22] The basic position that Hume took
towards the union was that England and Scotland should
enter the union as equal parties and that in the creation of the
union the best features of each nation should be drawn upon.
In keeping with this principle, Hume called for the adoption
of the Scottish form of church government in England.[23] The
proposal did not sit well either with the English ecclesiastical
and political establishment or with the Scottish episcopalians,
and it led to the suppression of Hume's second union treatise
abroad in 1610. It appeared to at least one hostile Scottish
reader that the tract pretended 'the union of Scotland with
England has no other end than to make Scotland equal to
England in all and superior in some points'.[24] The potential
superiority that was feared consisted mainly in the ecclesiastical superiority that would result from the establishment of a
British presbyterianism.

Hume was a lonely Scottish voice in 1605, when he wrote
the second treatise, and since the presbyterian movement had
little strength in England at this time, his proposal won no

---

[21] Williamson, *Scottish National Consciousness*, 20–47.

[22] D. Hume, *De Unione Insuale Britanniae Tractatus 1* (London, 1605); 'De Unione . . .
Tractatus Secundus', NLS Advocates MS 31. 6. 12; Edinburgh University Library,
Laing MS III. 249. On earlier desires for a British presbyterian church see G.
Donaldson, 'Foundations of Anglo-Scottish Union', in *Elizabethan Government and
Society*, ed. S. T. Bindoff *et. al.* (London, 1961), 304.

[23] D. Calderwood, *The History of the Kirk of Scotland*, ed. Thomson (Wodrow Society,
1842–9), vi. 731.   [24] SP 14/57/104.

apparent support there. Ridpath may have been right when he claimed that the English puritans favoured union in the first decade of the seventeenth century, but there is no evidence that they subscribed to Hume's peculiar type of unionism.[25] The idea of exporting presbyterianism lived on in Scotland, however, and in the 1640s it found much more forceful expression in the writings and formal demands of the Covenanters. Indeed, it became the centre-piece of Scottish demands in the negotiations with England between 1640 and 1647. Because the English puritans achieved a temporary ascendancy during those years, because the English episcopalian church lay in ruins, and because the parliamentarians in the Civil War (and eventually the royalists as well) required Scottish military assistance, there was some hope that the presbyterian plan would in fact be implemented.

The goal of the Covenanters was bold and comprehensive: the establishment of the true reformed religion in doctrine, form of worship, and church government in the three kingdoms of Scotland, England, and Ireland. Among these nations there was to be the 'nearest conjunction and uniformity' in all ecclesiastical matters.[26] The plan was predicated upon the presbyterian version of the Knoxian statement that the Scottish church was the purest and most completely reformed of all Protestant churches and hence the model upon which British uniformity of religion was to be built.[27] It was therefore a genuine expression of the Scottish imperial viewpoint. Like other Scottish appeals for religious union, it also accepted religious unity as an end in itself. It is true that the Covenanters argued for their plan on the grounds that it would provide security for their kingdom as well as their

[25] [Ridpath], *Discourse*, 44.
[26] *The Stuart Constitution 1603–1688*, ed. J. P. Kenyon (Cambridge, 1966), 264; *Some Papers of the Commissioners of Scotland Given in Lately to the Houses of Parliament concerning the Proposition of Peace* (London, 1646), 17–18. See also *A Declaration and Brotherly Exhortation of the General Assembly of the Church of Scotland to their Brethren of England* (London, 1647), 4; *A Brotherly Exhortation from the General Assembly of the Church of Scotland to their Brethren of England* (Edinburgh, 1649), 5.
[27] J. Knox, 'The Fourt Book of the Progresse and Continuance of Treu Religioun within Scotland' in *Works*, ed. D. Laing (Edinburgh, 1846–64), ii. 263–4; [D. Calderwood], *Quaeres concerning the State of the Church of Scotland* (n.p. 1638), 3; *A Short Relation of the State of the Kirk of Scotland since the Reformation of Religion* (Edinburgh, 1638), sig. A2; Burrell, 'Covenant Idea', 344.

church, bring lasting peace to Britain, and lay the foundation of a closer, more general union between the English and Scottish peoples.[28] They hoped that the 'mutual covenant' which they proposed would prevent the Scots from being viewed as strangers in England.[29] These non-religious reasons may help to explain why some lay Covenanters supported the demand for religious union.[30] But the religious objectives of the Covenanters appear to have been uppermost in their minds. Of the eight reasons they gave for their policy of seeking ecclesiastical uniformity with England, all but two were religious in nature.[31]

The imperial presbyterian vision of religious unity was not realized in the mid-seventeenth century, but it persisted as an ideal, surfacing again in 1660 and appearing in its final form in the late seventeenth and early eighteenth centuries.[32] At that time it found expression, oddly enough, in a number of pamphlets that were written against the proposed Treaty of 1707. The main argument of these pamphlets was that by providing security for the English church the Treaty prevented Scots from pursuing their imperialistic goal of reforming that church in accordance with the Covenant. Instead of objecting to the Treaty because it did not provide sufficient guarantees of the independence of the Scottish church, these pamphlets opposed it because it threatened their goal of achieving religious union.[33] As the author of *The*

---

[28] *The Scots Commissioners, their Desires concerning Unitie in Religion* (London, 1641), 10–12. In 1642 the General Assembly asked '[W]hat hope can the Kingdom and Kirk of Scotland have of a firm and durable peace till prelacy, which hath been the main cause of their miseries and troubles first and last, be plucked up root and branch?' R. S. Paul, *The Assembly of the Lord: Politics and Religion in the Westminster Assembly and the 'Grand Debate'* (Edinburgh, 1985), 66–7. For the expression of the commissioners' general unionist objectives see SRO, PA 7/24, ff. 275–276.

[29] *A Short Declaration of the Kingdom of Scotland for Information and Satisfaction to their Brethren in England* (Edinburgh, 1644), 4.

[30] Stevenson, *Scottish Revolution*, 220–1.

[31] *The Scots Commissioners, their Desires*, 5–7.

[32] I. M. Green, *The Re-establishment of the Church of England, 1660–1663* (Oxford, 1978), 13–15. For the appearance of this idea in 18th-century popular culture see D. Miller, 'Presbyterianism and Modernization in Ulster', *Past & Present*, 80 (1978), 70–1.

[33] *Queries to the Presbyterian Noblemen and Gentlemen, Barons, Burgesses, Ministers and Commoners in Scotland who are for the Scheme of an Incorporating Union with England* (Edinburgh, 1706); [J. Webster], *Lawful Prejudices against an Incorporating Union with England* (Edinburgh, 1707), 5

Smoaking Flax Unquenchable wrote, 'It is intolerable that Union can be where religion is not the object and end.'[34]

Despite their many differences, all religious unionists—English and Scottish, episcopalian and presbyterian—argued that the implementation of their programme would encourage the growth of political stability in Britain and strengthen the Protestant cause. The events of the 1640s, when religious differences between the two kingdoms, as well as within them, led to military conflict, seemed to prove them right. Religious unity also promised to strengthen the Protestant interest not only in international diplomacy but in domestic campaigns against Roman Catholics. It is not surprising that in his efforts to promote ecclesiastical union by strengthening episcopacy in Scotland, the Earl of Dunbar devoted a considerable amount of attention to the development of methods to check the spread of popery.[35] Somewhat ironically, a policy of incorporating union might actually have benefited Scottish Catholics, since English recusancy laws penalized Catholics less severely than did Scottish laws. In 1604 Alexander Macquhirrie, SJ, thought that if the union were accomplished, it would 'give us a great accession in numbers' for precisely that reason.[36] Most Englishmen and Scots, however, were not so finely attuned to the nuances of union and therefore continued to see anti-popery as a natural corollary of church union, keeping in mind the well-known opposition of the Jesuit, Robert Parsons, to Anglo-Scottish union.[37] It made perfect sense that a united Protestantism would serve as the best defence against popery and all other threats to the Protestant cause.

II

However laudable the goal of religious union appeared to be in the eyes of its various advocates, the task of implementing it was fraught with insurmountable difficulties. At the root of all

[34] *The Smoaking Flax Unquenchable: Where the Union betwixt the two Kingdoms is Dissecated, Anatomized, Confuted and Annulled* (n.p., 1706), 7.

[35] *Dictionary of National Biography*, v. 1125; *CSP Ven. 1607–10*, 312–13.

[36] *Spain and the Jacobean Catholics*, vol. i: *1603–1612*, ed. A. J. Loomie, SJ (Catholic Record Society, 1973), 182 n.

[37] [R. Parsons], *A Conference about the Next Succession to the Crowne of Ingland* (n.p., 1594), Pt. II, 118–23. See also *CSPD 1603–10*, 246.

these problems was the fact that the religious differences between the two countries were much greater than most contemporaries were willing to admit. To be sure, the common Protestant bond that they shared was a real one, strong enough to lay the foundation of a diplomatic alliance in the late sixteenth century and to support a common distaste for Roman Catholicism at home and abroad in the seventeenth.[38] But religious union required far more than a common anti-Catholicism. It demanded, as necessary pre-conditions, both a close similarity of doctrine, ceremonies, and church government and a compromising, ecumenical attitude on both sides. Neither of these pre-conditions obtained in 1603 or at any other time in the seventeenth century.

Unionists like Bacon and Cowper were in error when they claimed that as far as doctrine was concerned, the religious unity of England and Scotland was already 'perfect' at the Union of the Crowns.[39] The two churches were certainly closer in doctrine than in forms of worship or church government, but they did *not* agree on all fundamentals. One need only compare the Thirty-nine Articles of 1559 or even the Lambeth Articles of 1595 with the Scottish Confession of Faith of 1581 to appreciate the extent of the difference in this regard. As John Harington observed in 1602, Scotland was 'inclined to a purer manner of doctrine which they take from the reformed churches beyond the seas'.[40] Only briefly, during the 1640s, when the English Westminster Assembly adopted a Confession of Faith and the Scottish General Assembly accepted it, was there a sufficient doctrinal foundation for a genuine religious unity. Both in the 1630s and again after 1660 the doctrinal differences between the two churches widened rather than narrowed, as Anglicanism de-emphasized those Calvinistic elements in its creed that had previously provided some foundation for agreement with Scotland. The English apologist for union, Thomas Pyle, was conveniently overlooking these differences when he wrote in 1707 that 'the body and

---

[38] This bond often did not exist in the Borderlands, where the Catholic connections of the English commissioners prevented full co-operation with the Scots on the Border commissions. See L. C. Gay, 'The Border Commissions: Law, Politics and Faction in the Stuart Middle Shires', MA thesis (Univ. of Maryland, 1972), 90.

[39] *L&L* iii. 223 [W. Cowper], *The Bishop of Galloway His Dikaiologie* (London, 1614), 18.   [40] Harington, *A Tract on the Succession to the Crown*, 95–6.

substance is the same among us all; the fundamentals of our common Christianity and the vitals of the Protestant doctrine remain universally entire'.[41]

In matters of ritual the differences between the two countries were far greater than in doctrine. One indication of the gap between English and Scottish official practice is that the ecclesiastical ceremonies and customs which offended English puritans, such as the use of vestments, bowing at the name of Jesus, and using the cross in Baptism, had no place in the Scottish church. By the standards of English puritans, in other words, the Scottish church was fully reformed, whereas the English church was not. This religious difference between England and Scotland served as a constant impediment not only to ecclesiastical union but to other forms of union as well. Even during the reign of James I, when English puritans had much less to complain about than they did during the reign of his son, ceremonial differences hampered union initiatives. According to Lord Belhaven, writing in 1701, the 'great obstacle to an Union' in the early seventeenth century was precisely the conflict over religious rituals.[42] That obstacle became even more formidable during the 1630s and again in the 1660s, when the two churches grew further apart.[43]

Ceremonial differences, however profound they may have been, paled in comparison with those that prevailed in the structures of church government. As the Scottish commissioners observed in 1641, these differences were the 'main cause of all other differences between the two nations since the Reformation of religion'.[44] The English episcopalian system, which was essentially the Roman Catholic system of archbishops and bishops without the pope, contrasted dramatically with the Scottish presbyterian system of kirk sessions, presbyteries, provincial synods, and general assemblies. Aside from their denial of any papal power and their use of the territorial parish as the basic unit of ecclesiastical administration, the

---

[41] T. Pyle, *National Union a National Blessing* (London, 1707), 18.

[42] J. Hamilton, 2nd Lord Belhaven, *A Speech in Parliament on the 10. Day of January 1701 by the Lord Belhaven* (Edinburgh, 1701), 5.

[43] [Calderwood], *Certain Quaeres*, 5.

[44] *The Scots Commissioners, their Desires*, 8.

two systems exhibited few similarities. Both accommodated a clerical hierarchy, but the Scottish system, unlike the English, included lay elders and allowed for the democratic election of ministers. At the head of the English church stood the king, from whom all clerical power descended through the bishops and archdeacons to the rectors and vicars of the parishes. In the Scottish system power, which was exercised collectively at all levels, tended to flow upwards, and the king held no formal ecclesiastical power. It is true, of course, that from 1612 to 1637, and again from 1661 to 1689, Scotland had an episcopalian system that was grafted on to the basic presbyterian structure and to a greater or lesser extent coexisted with it. Presbyterianism, however, was never eliminated, and from 1592 to 1612, 1638 to 1661, and 1690 to 1707 it was the only established form of church government.[45] In England the episcopalian system survived throughout this entire period, except from 1646 to 1660, when first a presbyterian system, and then a congregational one, were established. It should be noted, moreover, that the English presbyterian system, which was never fully implemented, still differed significantly from its Scottish counterpart, mainly in that the English parliament exercised tight control over it. In Robert Baillie's famous words, the English presbyterian system of the 1640s was a 'lame, Erastian Presbytery'.[46]

All these differences between England and Scotland represented serious impediments to any plan for religious or ecclesiastical union. Whether these difference might have been resolved to the extent that one uniform and comprehensive church could have been established and accepted by the people of both nations is impossible to determine. Theoretically such unity was possible, but it would have required compromise on both sides, and neither church gave any indication that it was willing to make the necessary concessions. It is true that a spirit of compromise was not completely absent in either church during the seventeenth century. The English church had been built to some extent on the basis of

---

[45] On the structure of the Scottish church see W. R. Foster, *The Church Before the Covenants: The Church of Scotland, 1596–1638* (London, 1958).

[46] *The Letters and Journals of Robert Baillie*, ed. D. Laing (Edinburgh, 1841–2), ii. 362.

compromises between different religious viewpoints in the reign of Henry VIII and at the time of the Elizabethan Settlement, and some evidence of further compromise appeared at the Hampton Court Conference of 1604 and the Savoy Conference of 1661. In Scotland the coexistence of presbyterian and episcopalian forms of church government, especially during the 1610s and 1620s, reflected a similar spirit of accommodation. But when the question of Anglo-Scottish religious union arose, the spirit of compromise rarely manifested itself. Instead of making concessions to the other side in pursuit of a settlement acceptable to all, religious unionists usually spoke in terms of imposing their own particular system upon one country or the other, a tactic that virtually ensured a hostile and defensive reaction. Even the advocates of the Scottish presbyterian plan for church union in the 1640s, who at least claimed that they were open to the possibility of compromise, in fact left little room for such a development. Alexander Henderson and his colleagues do not deserve to be labelled 'proselytizing bigots',[47] but they had no intention of agreeing to establish a religious union that failed to meet their own exacting standards of 'reformation'.

Plans for the religious union of England and Scotland failed not only because they were impracticable and inflexible but because they encountered determined opposition in both countries. This opposition was probably more intense and widespread than opposition to any other single aspect of the union question, and it served as a major impediment to a union of any sort, not only in the early seventeenth but in the early eighteenth century as well.[48]

Opposition to religious union in England had a variety of sources. One of these was certainly the fear of Scottish presbyterian imperialism. Now one might question how seriously Englishmen took the threat of a Scottish-style church being imposed upon them. Surely England had the political and ecclesiastical resources to resist any such alteration of its traditional forms of worship and church government. Never-

---

[47] For a defence of Henderson against this charge see W. M Hetherington, *History of the Westminster Assembly of Divines*, 4th edn., ed. R. Williamson (Edinburgh, 1878), 118–19, 381; Ferguson, *Scotland's Relations with England*, 125.

[48] Pyle, *National Union*, 17; Mackenzie, *Parainesis Pacifica*, 7.

theless, the fear of such Scottish influences did in fact prevail at different times during the period 1603–1707, and it was not completely unfounded. It first appeared shortly after the Union of the Crowns, a time when Englishmen also expressed fears of Scottish political and legal influence in their country. Certainly there was an opportunity for Scottish ecclesiastical influence to be exercised. Just like his four English predecessors, James was expected to achieve a religious settlement at the beginning of his reign, and until that settlement was negotiated at the Hampton Court Conference in 1604, the position of James himself was unclear. It did not seem likely that he would favour the adoption of any sort of presbyterian form of church government (and of course he proved at the conference that he did not) but his Scottish nationality, his desire to reach an agreement with the puritans, and the enthusiastic welcome which moderate puritans gave him all caused consternation in conservative quarters.

After James allayed all these fears, the possibility of some sort of Scottish presbyterian influence did not arise again until the 1640s, when the collapse of the royal government, the necessity of gaining Scottish support for the parliamentary cause, and the ascendancy of the presbyterians in England gave the Scottish presbyterians a much greater chance of implementing their programme in England than they had had in 1603–4. Although abandoned by the parliamentarians, the programme was adopted by the royalists, who in their new-found alliance with the English presbyterians and the Scots made the introduction of presbyterianism on a trial basis one of the main planks in their political platform. All this agitation for religious union was 'so much preached and written against', mainly by the English Independents.[49] Their very real fear of Scottish influence extended to politics as well as religion, and in one pamphlet they referred emotively to Scottish plans to 'enslave us'.[50] After 1660 English fears of Scottish presbyterian influence abated but still appeared from time to time. In 1707 the Anglican minister Deuel Pead, confident though he was of his church's strength, still found it

---

[49] *Some Papers of the Commissioners of Scotland*, 18.
[50] *The Scottish Mist Dispel'd*, 14–15.

necessary to allay the fear that in a united Britain 'the Kirk of Scotland must over-run the Church of England'.[51]

The anti-Scottish sentiment that Pead referred to, as well as the earlier seventeenth-century fear of Scottish religious influence in England, probably reflected more a fear of collective Scottish clerical influence in England than a fear that the English form of worship or system of church government would be altered. This fear had a solid foundation in reality. Even before 1603 Scottish ministers had gained access to English benefices and pulpits,[52] and after the Union of the Crowns their numbers increased. Of thirty-two Scots ordained in the diocese of London between 1598 and 1628, all remained in England.[53] By English law those Scots who were born before 1603 and who were not subsequently naturalized needed a royal licence to hold a benefice, but even those who did not obtain such licences could serve as vicars and curates.[54] As naturalized Scots came of age, even this legal impediment became meaningless, so that during the period immediately preceding the Civil War Scots were believed to have possessed English benefices equal in value to all the benefices in Scotland.[55]

The fear that Scottish clerics would infiltrate and for all practical purposes overrun the English church found expression as early as 1604 in the work of Sir Henry Spelman. Recognizing that 'clergymen flow in Scotland and benefices ebb', Spelman believed that Scottish churchmen would be 'compelled by necessities at home to seek their fortunes abroad'. They would use their friends and countrymen to assist them, and since the Scottish clergy were almost all 'gentlemen of good houses', they would be able to use their kin and other connections to 'interlace the Scots with the English in all places'.[56] This type of clerical invasion of England constituted a much more plausible danger to the English

---

[51] D. Pead, *The Honour, Happiness and Safety of Union* (London, 1707), 12.

[52] P. Collinson, *The Elizabethan Puritan Movement* (London, 1967), 275–7; Donaldson, 'Foundations of Anglo-Scottish Union', 302–3.

[53] R. O'Day, *The English Clergy: The Emergence and the Consolidation of a Profession* (London, 1979), 3–4.

[54] 3 R. II, c. 3; 7 R. II. c. 12.

[55] P. Heylyn, *Aerius Redivivus* (London, 1672), 470.

[56] Spelman, 'Of the Union', in *Jacobean Union*, 175–6.

church than the appeals of men like David Hume and the Covenanters of the 1640s for the establishment of a presbyterian form of church government in England.

Spelman and those who shared his views were concerned not only with the loss of English benefices to the Scots but with the religious temperament of some of the clergy who would be travelling south. Spelman mentioned specifically that 'there is yet another fear touching Scotland, that whilst we receive their flowers and herbs so plentifully into our churches, we shall now and then shuffle in some of their weeds also. I mean those fiery spirited ministers that in the fury of their zeal have not only perverted the stable government of that church but even wounded the very kingdom itself.'[57] Spelman doubtless had in mind Scottish presbyterian ministers like Andrew and James Melville and John Davidson, the last of whom had earned the name of 'the thunderer' while preaching in London.[58] Although these men had counterparts within the English puritan community, they could easily strike English observers as a breed apart. After observing the conduct of the two Melvilles and James Balfour in their conference with King James in September 1606, James Montagu wrote to his brother that he had 'never heard such men nor such divinity since I was born'.[59]

To men such as Spelman and Montagu, as well as to many other Englishmen, the problem with Scottish ministers and the Scottish system of church government was mainly political. Spelman, as mentioned above, thought that the fiery-spirited ministers caused instability in the government of church and state, while Montagu observed that 'as for king and counsell both, they respect them not much'.[60] Certainly James VI and I, who had to contend with the Melvilles both in Scotland and in England, objected to these men and their system of church government for precisely these reasons. James, more than any one individual, popularized the phrase 'No bishop, no king', which neatly epitomized the threat that presbyterianism represented. The connection that he and

[57] Ibid., 176–7. See also the Earl of Northampton's reference to the 'giddiness of their overweening spirits' Winwood, *Memorials of Affairs of State*, ii. 94.
[58] Collinson, *Puritan Movement*, 277.
[59] Bodl., Carte MS 74, f. 379. I am grateful to Esther Cope for this reference.
[60] Ibid.

others established between presbyterianism and political instability became even more plausible, at least to Englishmen, after 1660, since they could readily interpret the rebellious activities of the Scots during the period 1638–47 as the work of presbyterians.[61] As late as 1702 the Englishman Blackerby Fairfax, writing for a Scottish audience and even pretending to be Scottish himself, boldy asserted that presbyterianism was inconsistent with the institution of monarchy and as dangerous to the Royal Supremacy as popery itself.[62] One year later the author of another virulently anti-presbyterian tract insisted not only that the presbyterians were anti-monarchical 'out of principle' but that their system of church government was too strong for the civil power and would result in the ruin of either or both.[63]

Some of the strongest English opposition to a presbyterian union came from the advocates of the alternative plan for an episcopalian union. One of the main attractions of reviving episcopacy in Scotland and uniting the two churches was that it would dampen the religious and political enthusiasm of the Scottish presbyterians. John Doddridge made this clear in 1604 when he recommended that the two countries embrace the form of church government that was 'farthest off from popular faction, most obedient to the ecclesiastical and civil magistrate, and least subject to mutability and fantastical opinions'.[64] Doddridge's statement is clearly unionist, but it constituted a strong argument against religious union on the Scottish presbyterian model.

English opposition to religious union was not usually inspired by nationalism. The defence of the English church against external threats did of course have the potential for exciting intense nationalistic sentiment, as the struggle against the papacy in the early sixteenth century had revealed. The threat of Scottish religious imperialism was not powerful enough, however, to galvanize a large section of the English nation to a defence of the liberty of its church. The

---

[61] *The True Copy of a Letter directed to the Provost and Preachers of the City of Edinburgh* (1600), in *Somers Tracts*, vii. 489–91.

[62] Fairfax, *Discourse upon the Uniting Scotland with England*, 28–9.

[63] *The State of Scotland under the Past and Present Administration* (n.p., 1703), 5–6.

[64] Doddridge, 'Breif Consideracion', in *Jacobean Union*, 149.

only nationalism that the entire question of religious union generated in England was the xenophobic reaction to the threatened invasion of Scottish ministers.

In Scotland the situation was quite different. The real possibility that episcopacy would become strong enough to destroy or emasculate the presbyterian system of Scottish church government led ministers and statesmen alike to equate the threat of an episcopalian union with a direct attack upon the independence and the integrity of the Scottish nation. The danger that the Scottish church would be absorbed into a larger British ecclesiastical establishment appeared to be just as serious and threatening as the danger that the Scottish parliament would suffer a similar fate. When Scottish anti-unionists wrote or spoke about the threat to their liberties, they almost always included their church, either explicitly or implicitly, among the national institutions that they saw as being faced with possible extinction. The fact that the Scottish church had served as a major source of national cohesion since the Reformation, filling a role that the weak and subservient Scottish parliament could not play, made the preservation of the church an issue of extreme national importance. This was especially true in the localities, where anti-unionist sentiment usually expressed itself in the pulpit or in the presbyteries, both at the beginning of the seventeenth century and just before the Treaty of 1707.[65]

Concern for the autonomy of the Scottish church did not always take the form of opposition to ecclesiastical union as such. On the one hand, those who pleaded for the maintenance of the integrity of the Scottish church often favoured a different, presbyterian type of unity. On the other hand, the advocates of Scottish ecclesiastical autonomy did not always describe the danger facing their church as its possible absorption into a united British ecclesiastical establishment. The much more immediate danger was an English, episcopalian attack upon the integrity of their Church. During the periods of 1603–38 and 1661–89 the main problem was the dilution of

---

[65] Hetherington, *History of the Church of Scotland*, 118–19; *The Humble Address of the Presbytry of Hamilton* (Edinburgh, 1706); *The Humble Address of the Presbyterie of Lanerk* [n.p. 1706]; *An Account of the Burning of the Articles of Union at Dumfries* (n.p., 1706); HMC. *Mar and Kellie MSS*, 335.

the powers of presbyteries by the reintroduction and strengthening of the Scottish episcopacy.

After 1689, when episcopacy was formally abolished and the powers of the monarchy *vis-à-vis* the church were weakened, the threat to the Scottish church took a different form: the potential danger of ecclesiastical legislation by a united British parliament. Many Scottish statesmen and clerics simply could not be persuaded that all the ecclesiastical guarantees which were included in proposals for union after 1689 would in fact be observed. These men were reluctant to entrust the protection of their national church to a British parliament that included twenty-six English bishops and an overwhelming majority of English members, and they realized that the same British parliament would have the power to overturn even those articles of the Treaty of Union which were declared to be fundamental. They foresaw the granting of toleration to Scottish episcopalians, noted that Scots would not be exempt from the Sacramental Test as a condition of holding public office in England, and objected that the members of their church would be obliged to take the oath abjuring a successor to the crown who was not a member of the Church of England.[66] These criticisms of the Treaty reflect the fear that 'prelacy, instead of being extirpate, shall be further rooted here',[67] leading ultimately to the dis-establishment of presbyterianism. Fletcher of Saltoun believed that the English felt duty-bound to 'demolish' a separate established church government in Scotland.[68]

Scottish suspicion of unlimited British parliamentary power over its church and the attendant fear of a re-establishment of episcopacy reflect another, perhaps deeper reason why many Scots were reluctant to accept any sort of plan that involved an actual assimilation of the two churches. The theory of Erastianism, which asserted the supremacy of the state over

---

[66] *The Grounds of the Present Danger of the Church of Scotland* [Edinburgh, 1707], 1–2; [R. Wyllie], *A Speech without Doors, concerning Toleration* (n.p., 1703); [W. Carstares], *The Scottish Toleration Argued: or, an account of all the laws about the Church of Scotland ratif'd by the Union-Act* (London, 1712).

[67] *Grounds of the Present Danger*, 2. See also Webster, *Lawful Prejudices against an Incorporating Union*, 11; [Symson], *Sir George M'Kenzie's Arguments*, 3; *A View from the North or an Answer to the Voice from the South* [Edinburgh 1707].

[68] Fletcher, *State of the Controversy*, 10.

the church, never attracted the same degree of support in Scotland that it did in England. In defining the relationship between church and state Scots tended to endorse some version of the doctrine of the two kingdoms—the theory, developed by Andrew Melville in the late sixteenth century, that civil and ecclesiastical authorities occupied two separate spheres of activity.[69] This theory did not deny the temporal magistrate all power in ecclesiastical matters. As the Scottish commissioners who negotiated with England in 1646 explained, parliament was 'superior to all the assemblies of the Church in place, dignity, honour and earthly power', and could compel ministers and assemblies to 'perform the duties which Christ requires of them'. Nevertheless its power was not 'formally ecclesiastical' and it did not possess 'a headship and supremacy in the church', as it did in England.[70]

As Scottish parliamentary power and prestige increased in the late seventeenth century, Scots assigned an increasingly powerful role to parliament in regulating the church, but they still insisted upon limitations to that power. The English minister and historian White Kennett may have been exaggerating, but he did have a point when he argued in 1702 that the Scots 'are utter enemies to all Erastian principles and will not suffer the spiritual power of the Kirk to depend precariously upon the state, as bagpipes do upon the blower of them'.[71] The most uncompromising attack upon Erastianism in general, and upon the English church in particular, appeared in a strongly anti-unionist pamphlet written in Scotland in 1706.[72] Even within the more moderate anti-unionist circles, opposition to Erastianism remained strong in the early eighteenth century. As mentioned above, Fletcher of Saltoun, in arguing for the preservation of a separate established church in Scotland, rejected the English Erastian belief that church and state were inextricably interwoven. For Fletcher at least, an established Scottish church was not to be

---

[69] See Burrell, 'Covenant Idea', 344–7.

[70] *Some Papers of the Commissioners of Scotland*, 19–20; *Correspondence of the Scots Commissioners in London, 1644–1646*, ed. H. W. Meikle (Edinburgh, 1917), p. xxv. See also the assault on Erastianism in *Letters and Journals of Robert Baillie*, ii. 360–2.

[71] [W. Kennett], *A Letter from the Borders of Scotland* (London, 1702), 7.

[72] *Smoaking Flax Unquenchable*, 17–19.

a 'state church' in the same way that the English church was.[73]
His fear, of course, was that the Scottish church would
become precisely that type of church if the Treaty of Union
were to be approved.

III

The strength of both English and Scottish opposition to the
various proposals advanced for religious union, the differences
between the English and Scottish churches, and the unwillingness of both sides to seek a compromise solution ensured that
this aspect of programmes for 'perfect union' would not be
realized. The prospect of such a union, however, died a slow
death. The attraction of the idea of religious union, the
commitment of certain parties on both sides of the Tweed to
bring it about, and the belief that a certain amount of religious
union had already been realized kept the idea alive during the
entire period 1603–1707. Only after many efforts had been
made to implement it and after much inconclusive debate was
the idea completely abandoned.

The history of these efforts to achieve religious union is long
and complex. The period immediately following the Union of
the Crowns was one of great uncertainty in this regard, but
King James soon made it clear that he envisaged a union of
the two churches upon an episcopalian model, and in so doing
he established the frame of reference in which the issue was to
be debated throughout his reign and most of that of his son.[74]
James did not, however, try to implement his programme with
undue haste. The proposals he submitted to the commissioners for the union in September 1604 included a provision for
preserving the fundamental ecclesiastical (and civil) laws
'entire',[75] and if the king did in fact give serious consideration
to appointing Richard Bancroft as primate of Great Britain
rather than primate of England, he decided in the end against
making such a provocative move.[76] As it was, the question of
church union did not become the subject of royal policy until

---

[73] Fletcher, *State of the Controversy*, 10–11.
[74] For foreign opinion on James's plan see Bodl. Calendar of Carte MSS, vol. 220,
31 May 1606.
[75] SP 14/9A/35.
[76] *CSP Ven. 1603–7*, 201.

after most of the other features of the Jacobean union project failed to win parliamentary approval. It should also be noted that James never actually attempted to assimilate the two churches. Instead he contented himself with achieving as much conformity between them as possible in doctrine, liturgy, and form of church government.[77]

These efforts proved to be fairly successful. With respect to doctrine the Jacobean innovations did not amount to very much, but one must remember that it was in this area of religion that the differences between the two churches were least pronounced, and that James throughout his life adhered to most of the fundamentals of Calvinism. A new confession of faith approved by the General Assembly at Aberdeen in 1616 brought the Scottish church a little more into line with the English by reversing earlier official Scottish condemnations of certain points of doctrine as being superstitious.[78] The changes with respect to liturgy, embodied mainly in the Five Articles of Perth in 1617, represented a much more substantial challenge to traditional Scottish practice. The re-establishment of the sacrament of confirmation, which was to be performed by bishops, did not amount to very much simply because it was rarely performed. But the reintroduction of five holy days, especially Christmas and Easter, and the infamous fifth article, which required kneeling at the reception of communion, appeared to be distinctly Anglican, if not popish, moves.

In the area of church government James made the most significant changes of all. Through a series of steps taken in parliament, in the General Assembly, and by his own prerogative power between 1597 and 1612, James revived the practically moribund institution of episcopacy in Scotland. He secured the admission of the bishops to parliament, the restoration of their estates and income, their appointment as moderators of synods, and their right to appoint judicial subordinates, to present ministers to benefices, and to exercise ecclesiastical jurisdiction.[79] Although Scottish bishops did not possess as much power as their English counterparts, and

---

[77] Scottish ministers claimed that James repeatedly promised not to bring about uniformity between his two churches. Lee, *Government by Pen*, 167.

[78] Ibid. 160.

[79] Foster, *Church Before the Covenants*, 15–32.

although the episcopalian system over which they presided coexisted with a presbyterian structure that was never discredited, James did bring the English and Scottish systems of church government into closer conformity. The similarities became even more apparent in 1610, when two Scottish courts of High Commission, modelled after those in England, were established.[80]

Whether James could have actually implemented a full-scale ecclesiastical union remains a matter of speculation. Although the king possessed significant prerogative powers in ecclesiastical affairs in both countries, and although James was not averse to exercising these powers when necessary, he did not have sufficient constitutional authority to join the churches by his power alone. Even if he did, James would have recognized the political necessity of gaining approval for such a plan from the two parliaments, the English convocations, and the Scottish General Assembly. In all these institutions James would have encountered serious opposition, similar in nature to the general anti-unionism that had arisen in both countries between 1604 and 1607. Even in the English convocations, which naturally approved of his Scottish ecclesiastical policies, the prospect of actual church union would have excited xenophobic fears of Scottish infiltration of the clerical establishment. So even if James had hoped to realize this one aspect of his original union project, he did not press the issue and therefore had to content himself with a partial conformity rather than uniformity, much less union. Perhaps the best example of James's restraint in this regard was his reluctance to impose an English prayer-book on Scotland at the end of his reign.[81]

As it was, the pursuit of partial conformity had created a considerable amount of ill will, at least in Scotland. Each of the measures he had taken to bring the Scottish church more into line with that of England had encountered a certain

[80] Lee, *Government by the Pen*, 96. It is uncertain whether this move was specifically intended to bring about greater conformity between the two churches. G. I. R. McMahon, 'The Scottish Courts of High Commission, 1610–1638', *Records of the Scottish Church History Society*, 15. 3 (1965), 196.

[81] Ferguson, *Scotland's Relations with England*, 108. James also did not press for the uniformity of clerical costume. See C. V. Wedgwood, 'Anglo-Scottish Relations, 1603–1640', *TRHS*, 4th ser. 32 (1950), 36.

measure of opposition from the General Assembly. Most of this resistance can be attributed to presbyterian abhorrence of prelacy, but it also sprang from two other sources. The first of these was an interpretation of the king's policies as being absolutistic. Of course the very tactics that the king used to revive episcopacy—the use of the prerogative in 1600, the reliance upon the Committee of Articles to facilitate parliamentary approval of his policies, and the manipulation of clerical votes in the General Assembly—support this interpretation. Even more important, however, the revival of episcopacy itself gave the king a new instrument of absolutism.[82] As many commentators later observed, the king, under the influence of the high church party in England, endeavoured to erect an absolutism in the church as well as the state.[83] The second source of resistance was the justified feeling that the king had reneged on his promise of 1604 to preserve the fundamental ecclesiastical laws of Scotland.

If James's Scottish eccelesiastical policies provoked determined opposition, those of his son eventually led to rebellion. Charles I not only perpetuated James's policies but carried them further than James had ever thought it politic to do. Whereas James was content to settle for partial conformity between England and Scotland, Charles and Archbishop Laud pushed for complete uniformity and in some respects actual union. The Scottish canons imposed in 1636 foreshadowed the infamous English canons of 1640, while the Scottish prayer-book of 1637, which was drafted in collaboration with the Scottish bishops, went a long way towards bringing Scotland into close liturgical conformity with England.[84] The inclusion of Scottish members on the English commissions for ecclesiastical causes of 1629 and 1633 represented another, more tentative move towards the union of the two churches. Scotland, nevertheless, continued to come under the jurisdiction of its own commissions, the last of

---

[82] Foster, *Church before the Covenants*, 30–1.

[83] *A Short Relation of the State of the Kirk of Scotland Since the Reformation*, sig. A2ᵛ; Ridpath, *Discourse*, 43–4.

[84] Laud and Charles originally intended to impose the English prayer book, but the Scottish bishops suggested changes. See M. Lee, Jr., *The Road to Revolution: Scotland under Charles I, 1625–1637* (Urbana, Ill., 1985), 204–5.

which was constituted in 1634.[85] On a more informal basis, Archbishop Laud assumed the function of 'overseer' of ecclesiastical affairs in Scotland, while it came to be believed that his ultimate goal was 'to bring the three churches of Scotland, England and Ireland to be governed by one Metropolitan'.[86]

The Scottish ecclesiastical policies of Charles I did, of course, involve much more than an attempt to achieve religious union. Much of the controversy that surrounded the implementation of these policies centred on their 'Arminian' or popish character and on Laud's exclusive reliance upon royal authority to introduce them.[87] Nevertheless, the widespread assumption in Scotland that the Caroline and Laudian measures were intended ultimately to bring about an amalgamation of the two churches, and possibly even a political union, did contribute in a significant way to the opposition they stimulated.[88] The Covenant signed in 1638 was, after all, a national covenant, and it reflected the same type of nationalistic determination to protect the integrity of Scottish institutions (political as well as ecclesiastical) as later found expression in the writings and speeches of men like George Ridpath and Fletcher of Saltoun. And this nationalist sentiment, while not unequivocally anti-English or anti-unionist, was founded upon opposition to a particular type of English policy and a particular type of unionism. It is no coincidence that the Covenant appealed in the most specific terms to the defence of the laws and liberties of Scotland included in the Scottish Act of Union of 1604.[89]

What the National Covenant did, in fact, was to change the

[85] R. G. Usher, *The Rise and Fall of the High Commission* (Oxford, 1913), 250; SP 16/131/27–9; *Foedera, Conventiones, Literae et . . . Acta Publica*, ed. T. Rymer (10 vols; The Hague, 1739–45), viii, pt. IV 35; McMahon, 'Scottish Courts of High Commission', 198. Scottish High Commissions, unlike their English counterparts, did not have a statutory foundation. See [D. Calderwood], *The Altar of Damascus or the Pattern of the English Hierarchie and Church Policie obtruded upon the Church of Scotland* ([Amsterdam, 1621]), 38.

[86] McMahon, 'Scottish Courts of High Commission', 208; *The Hamilton Papers*, ed. S. R. Gardiner (Camden Soc., 1880), 259.

[87] *A Short Relation*, sig. C3ᵛ, charges that Laud's innovations were 'not warranted by law'. For the Scottish charges against Laud in 1640 see J. Rushworth, *Historical Collections* (London, 1721), iv. 113–18.

[88] [Calderwood], *Certain Quaeres*, 8–9; Wedgwood, 'Anglo-Scottish relations', 32.

[89] Williamson, *Scottish National Consciousness*, 140–6.

frame of reference within which the question of union was being discussed. While rejecting emphatically the episcopal and increasingly absolutistic form that union was threatening to take, it laid the foundation for an alternative approach. Drawing upon the alliance between Scottish ministers and English puritans that harked back to the Elizabethan period, and tapping the qualified but nonetheless genuine unionist sentiment that characterized the work of men like David Hume, the Covenant set forth a radically different conception of religious and even non-religious union than had been countenanced during the previous four decades.

The first efforts to formulate this policy occurred in the Assembly at Edinburgh in 1640, and the agreements reached there laid the basis of all subsequent Scottish negotiations with the English parliament, and then with Charles I, during the 1640s.[90] Because of the English need for military assistance and because of the destruction of the entire episcopalian structure, there was in fact a greater likelihood of a religious union being achieved during these years than at any other time in the seventeenth century. To this likelihood the Solemn League and Covenant, the Westminster Confession of Faith, and the approval of an English presbyterian system of church government all bear eloquent testimony. The prospect of such unity, however, was not to be realized, both because presbyterianism did not have widespread support and because the English presbyterians were too Erastian for their Scottish co-religionists. The rise of the English Independents, who were feared and abhorred in Scotland, sealed the fate of presbyterianism in England, driving the Scottish presbyterians into a temporary alliance with the royalists and ending, for all practical purposes, the last real hope of religious union in the history of the two nations. Religion, which in the 1640s constituted the main source of unionist sentiment in both kingdoms, had by 1646 become the main source of disagreement between them.[91]

During the 1650s, when Scotland came under the direct military and political control of an England dominated by

[90] D. Stevenson, 'The Radical Party in the Kirk, 1637–45, *Journal of Ecclesiastical History*, 25 (1974), 149.

[91] *Correspondence of the Scots Commissioners*, ed. Meikle, pp. xxii–xxv.

Independents, the history of both churches and the relationship between them entered a new phase. During this time
both national churches were deprived of their formal clerical
assemblies. In England the two convocations and the legislatively authorized presbyterian national synod, which had
never been convened, were destroyed in rapid succession,
while in Scotland the General Assembly did not meet after
1653. Neither country, moreover, acquired new articles of
religion or a new liturgy. Instead of enforcing religious conformity, the government granted a broad toleration to the
subjects of both England and Scotland. This policy of toleration was anathema to Scottish presbyterians, but under the
circumstances it had its benefits, for it meant that they could
practise their religion and govern their own local churches
with little interference from England. Scottish presbyterians
may have regretted the end of the hopes they had for church
union in the 1640s, but at least they did not have to deal with
an ecclesiastical union imposed by England, which
Cromwell's control of Scotland and the establishment of a
united British parliament in 1654 had made technically
possible. The only type of religious union that might have
emerged in the 1650s was that which the English nonconform-
ist John Humfrey proposed a half-century later. Humphrey
envisaged a loosely structured national British church in
which the various established and tolerated churches would
accept their subordination to the reigning monarch.[92]

Programmes for religious union once again received serious
consideration after the Restoration, but they failed to attract
the same type of enthusiastic government support that they
had received in the first part of the century. During the 1660s
proposals for church union took two radically different forms,
each of which corresponded to one of the earlier plans. The
first was the programme of Scottish Protesters, English
nonconformists, and Ulster presbyterians for establishing
a 'close correspondence' of religion in all three kingdoms.[93]
This plan, which represented a variation on the Scottish
presbyterian theme, was short-circuited in 1661 by the re-

[92] Humfrey, *A Draught for a National Church Accommodation.* See also Chamberlen,
Great Advantages, 30.
[93] Green, *Re-establishment of the Church of England,* 13–15.

establishment of episcopacy in all three countries. The second form was the episcopalian scheme of Charles II, who effected very similar church settlements in England and Scotland. His goal was apparently to achieve conformity rather than actual church union, but even in this limited objective he realized less success than his grandfather. The main problem, as William Ferguson has pointed out, is that whereas episcopacy was re-established in England on a solid parliamentary and popular foundation, the episcopalian experiment in Scotland 'suffered shipwreck'.[94] Even if episcopacy attracted more support in Scotland than most historians have been willing to concede, it met with much more determined, systematic, and widespread opposition than it had during the period 1600–30, and it had to contend with an alternative presbyterian structure of church government with which it found it increasingly difficult to coexist.[95] The king's assumption of virtually absolute control over the Scottish church by virtue of the Act of Supremacy, which technically could have facilitated a plan for church union, in fact weakened its prospects, and in a real sense weakened the cause of political union as well.[96]

The problem with Charles II's plan for ecclesiastical union ran deeper than the unequal strength of the two episcopal establishments. In many ways the two churches diverged after 1660. The Restoration may not have signalled an unqualified victory of the Laudians in England, but the church that was re-established there developed a distinctly high-church, anti-puritan character, especially after the Clarendon Code created an official policy against nonconformists.[97] The differences between that post-Restoration church and the genuinely Calvinistic church of the Jacobean period, which had accommodated many moderate puritans, especially during the archiepiscopate of George Abbot, was quite pronounced.

[94] Ferguson, *Scotland's Relations with England*, 146.

[95] M. Lee, Jr., 'Comment on the Restoration of the Scottish Episcopacy, 1660–1661', *Journal of British Studies*, 1 (1962), 52–3; I. B. Cowan, *The Scottish Covenanters, 1660–1688* (London, 1976), Foster, *Bishop and Presbytery*, 11.

[96] *APS* vii. 554; Lee, *The Cabal*, 62.

[97] On the strength of Laudianism in England after 1660 see Ferguson, *Scotland's Relations with England*, 145–6; Green, *Re-establishment of the Church of England*, 22–4; R. Hutton, *The Restoration: A Political and Religious History of England and Wales, 1658–1667* (Oxford, 1985), 177–8.

The latter had some potential for union with the Scottish church, even if that union was never realized. The English church of the Restoration, however, did not have that potential, especially since the Scottish church did not deviate from its traditional Calvinistic orientation. One need only read a sample of the presbyterian, anti-English tracts of the early 1700s to appreciate the enormous doctrinal as well as ecclesiastical gulf that developed between the two churches after 1600.[98] The gulf was so wide that the king's government, despite its professions of support for the ideal of religious union, did not even attempt to enforce the Five Articles of Perth or a compulsory prayer-book in Scotland.[99] Just as English law and Scots law grew apart during the Restoration period, so too did the churches of the two countries.

The cause of religious union suffered from another development that occurred in England between 1660 and 1689: the erosion and eventual destruction of the concept of a comprehensive church. The idea that all Englishmen belonged to the same church, a national church, had been the guiding principle of English ecclesiastical policy since the Reformation. It had been one of the main assumptions upon which the Henrician, Edwardian, and Elizabethan settlements of religion had been based, and it had led to the treatment of nonconformists, even during the Laudian period, as fallen brethren who were to be reunited to the communion rather than as members of another sect outside the fold. The puritans themselves, in their commitment to reform the church from within rather than to separate from it, accepted the principle of a comprehensive church. Even during the Cromwellian period, when the state allowed a broad toleration, the commitment of the Independents to some sort of comprehensive national church, regulated by parliament, did not disappear. After the Restoration, however, the attitude of the political and ecclesiastical establishment began to change. By inaugurating a systematic programme of persecuting nonconformists while paying only lip-service to the ideal of reuniting them to the church, the Cavalier Parliament implicitly recognized the existence of churches other than

---

[98] See for example *Smoaking Flax Unquenchable*, 17–22.
[99] Cowan, *The Covenanters*, 49.

those established by law. When the state exchanged this policy of persecution for one of toleration (by prerogative in 1672 and 1687 and by parliamentary statute in 1689), the official policy of comprehension came to an end.

The effects of this change, which had no parallel in Scotland in 1689, had profound implications for the prospects of religious union, for a policy of Anglo-Scottish church union was predicated upon the achievement of ecclesiastical unity in each country. Indeed, both policies were pursued concomitantly during the first half of the seventeenth century, when the term 'church union' could refer to either national or British religious unity, or to both. Once the English abandoned the cause of ecclesiastical unity within their own country, there was little reason to pursue it within a larger and religiously more complex and varied Britain.

After 1689 unionists still occasionally spoke in terms of a vague Protestant bond between Englishmen and Scots that would help to fuse them into one nation at home and provide a united front against Catholicism abroad. Some pamphleteers, such as John Humfrey and Hugh Chamberlen, even spoke of a comprehensive 'national religion' in which a broad diversity of churches would be tolerated. In such an arrangement the government of the church would have few coercive powers. Chamberlen claimed that 'the only thing in religion deserving punishment' would be 'a breach of the civil peace',[100] whereas a similar and uncharacteristically Scottish appeal for a comprehensive British church government by Colonel John Buchan of Cairnbulg promised perpetual toleration to all whose principles and practices were 'consistent with the civil security'.[101]

None of these proposals for a tolerant national church attracted very much support, and even Chamberlen himself doubted whether there was 'any likelihood of success'. The problem was that however laudable the ideal, there was little practical reason for establishing the outward forms of a national church government that served no apparent religious or political function. A national church such as that envisaged by Chamberlen, Humphrey, or Buchan was little more

[100] Chamberlen, *Great Advantages*, 30.
[101] [Buchan] *A Memorial*, 7.

than a symbol of national unity. Certainly it was in no way essential to the broader cause of political and economic union with which most English unionists, and even Chamberlen and Buchan themselves, were concerned. The fact remains that with the establishment of toleration in England in 1689, English statesmen and MPs had determined that membership of all the citizens of one state in the same church was not essential to the political stability of the state. The same conclusion applied just as much, or even more, to the proposed British state. The cause of Anglo-Scottish religious union, in other words, lost the political rationale that it once had, at least from the English point of view, and once that happened, it no longer appeared to be a necessary precondition or corollary of political union.

The net effect of all these religious, ecclesiastical, and ideological changes after the Restoration was the virtual elimination of religious union from the broader question of political union. Unsuccessful in the past, unlikely in the future, and a source of more sentiment against the union than for it, the issue was not even submitted for discussion in the union negotiations of 1670, 1702, or 1706. Nor did the government of either country pursue the goal as an independent or parallel policy, in the manner of James I or Charles I. Instead of declaring their support for what was once widely believed to be a salutary goal, the statesmen of both countries who favoured a union either remained silent on the issue or took pains to assure the people of both countries that political union would *not* involve a union of churches. Eventually, in order to secure ratification of the union it became necessary to amend the Treaty with the two Acts of Security, which defended the established church of each nation from the allegedly proselytizing and imperial plans of church union by the other.

For some Scots, especially the Cameronians, the security thus provided was not adequate, and they waged a campaign against the Treaty, both before and after its ratification, on precisely these grounds. Aware of the omnicompetence and sovereignty of the British parliament, they feared that parliament would weaken the security it had given to the Scottish church by granting toleration to the Scottish episcopalians, an apprehension that was sustained in 1712. They also feared that the British parliament might repeal the Act of Security itself, even though it had declared that act to be fundamental.[102] That fear, as well as the general fear that the British parliament wished to destroy the independence of the Scottish presbyterian church and possibly even bring about religious union with England, proved to be unfounded. The British parliament, like the English commissioners who negotiated the Union, was willing to let the Scottish church be. Parliament did demand, not without cause, that a toleration should be established in Scotland to parallel that of England, but that demand did not imply, any more in Scotland than in England, that the state church would thereby be weakened.

Sentiment in favour of religious union did not completely die out in Scotland or England after the Treaty of Union was ratified. Committed Scottish presbyterians still spoke about their covenanted obligation to reform episcopacy in England, while English ministers like John Ollyffe regretted that the union would not be 'perfect'.[103] But for practical reasons it would have been difficult for either group to press its case. Religious pluralism, both within each country and within Britain as a whole, was legally established by 1712, and the religious case against it lost strength at the same time that its political benefits became readily obvious.

What happened, in sum, between 1603 and 1707 was that the union, considered both as an ideal and a reality, had been secularized. In 1603, and indeed long before that date, statesmen could not conceive of a union that did not have a religious dimension. All of the advocates of further union in the early seventeenth century recognized the necessity of some sort of religious unity. They may have had different ideas of what that union entailed, but none could deny John Doddridge's conclusion that without religious union there could be no perfect union.[104] By 1707 virtually all the advocates of an incorporating union, either explicitly or implicitly, denied that conclusion. Adherence to the same

[102] Webster, *Lawful Prejudices against an Incorporating Union*, 9.
[103] *Queries to the Presbyterian Noblemen*; John Ollyffe, *A Sermon Preach'd May the 4th 1707* (London, 1707), 19.
[104] Doddridge, 'Breif Consideracion', in *Jacobean Union*, 149.

doctrine, worship according to the same liturgy and, most emphatically, subordination to the same type of church government simply were not prerequisites for membership in the same state. Even a vague common 'Protestantism', which Defoe continued to celebrate in 1706, was not necessary.[105] The union was no longer to signify two nations, like the tribes of Israel, going up to Jerusalem together, but merely two nations sending representatives to a common parliament. Instead of making 'one entire body in one Covenant', the union would fashion only one body politic.[106]

Both a cause and an effect of this secularization of the union was the secularization of the state itself. Once the English state ceased enforcing conformity to one established religion, membership in one state no longer implied membership in one church, and this change was naturally reflected in the creation of a larger British state. Once this larger, secularized British state was created, it in turn helped to advance this process of secularization in two ways. First, it secularized Scottish political society, which until 1707 remained generally committed to the principle of enforced religious conformity and to the idea of the 'godly state'.[107] After 1707 Scots were forced to recognize the reality of ecclesiastical pluralism within the broader British state of which they were now a part, and after 1712 they had to recognize it within the more narrow bounds of Scotland itself. Second, the existence of a large, secularized British state helped to clarify the nature of a 'state church'. Between 1689 and 1707 Englishmen, while accepting the reality of toleration, none the less also recognized that they had an established church which still could be viewed as an arm of the state. After 1707, however, Englishmen had to accept the fact that there were two established churches in one state, and this unusual arrangement reduced the civic character of the Anglican church.

The British state established in 1707 was therefore something quite different from the model of a British unitary state that statesmen like James VI and I appealed to and

[105] Defoe, *Essay at Removing National Prejudices*, Pt. 1, 22–4.
[106] *A Declaration and Brotherly Exhortation*, 1.
[107] C. Larner, *Enemies of God: The Witch-Hunt in Scotland* (Baltimore, Md., 1981), 199.

hoped to establish at the beginning of the seventeenth century. To the extent that it had a united parliament and was not a federal state it was of course 'unitary' in the sense which that work carries today. But it did not conform to the early seventeenth-century concept of a unitary state, which was inspired by the sixteenth-century English example. That unitary state, in addition to having a single parliament and administration, had a legal unity (the common law) and an ecclesiastical unity (the English church). As we have seen, the British state erected in 1707 had neither of these attributes of unity, and therefore it represented a different conception of the state than that which Englishmen and, to a lesser but real sense, even Scotsmen had grown accustomed.

5

Economic Union

THE commercial and fiscal provisions of the Union Treaty of
1707 stand in dramatic contrast to its legal and religious terms.
Whereas the Treaty accepted a legal and religious pluralism
within a united British state, it laid the foundation for the
complete integration of the English and Scottish economies. It
turned Great Britain into the largest free-trade zone in Europe
and entrusted the regulation of both Scottish and English
trade and industry to the British parliament. Economically,
the British state established in 1707 was almost as unitary in
character as the English state before the union. The treaty
gave all Scots and Englishmen the 'full freedom and intercourse of trade and navigation' to and from any port or place
within Britain or its colonies, and it decreed that all Scottish
ships, like their English counterparts, would pass as British
ships. It provided that 'all parts of the United Kingdom . . .
shall have the same allowances, encouragements and
drawbacks, and be under the same prohibitions, restrictions
and regulations of trade, and liable to the same customs and
duties on import and export'. The Treaty also established a
standard value for coinage throughout Britain and applied the
same rules to the Scottish and English mints. In the most
comprehensive article of all, it established an identity of
English and Scottish law regarding trade, the customs, and
the excise.[1]

The Treaty did, to be sure, make a number of concessions to
Scottish economic interests. It exempted Scotland from duties
on certain products, in most cases for a period of seven years,
and it compensated Scotland for the portion of the higher
customs and excise duties that it would be paying towards the
relief of the English national debt. The means of achieving this

[1] Pryde, *Treaty of Union*, 84–98 (Articles 4–18.)

compensation, the 'Equivalent', recompensed those Scots who would suffer a financial loss because of the standardization of the coinage as well as those who had already lost by investing in the Darien scheme. Some of the Equivalent was also slated to pay the debts of the Scottish crown and to encourage Scottish manufactures. None of these concessions, however, gave Scotland a degree of economic autonomy within the union. Nor did the establishment of a Scottish court of Exchequer or the preservation of both the Bank of Scotland and the Scottish mint allow Scotland to govern its own economic life to any appreciable degree. The only way to secure a measure of economic independence for Scotland would have been to place limitations on the British parliament's authority to regulate trade and industry. This the Treaty did not do.

The fact that the Treaty of Union made provision for an economic union should not surprise us. From the very earliest discussions of further union at the beginning of the seventeenth century until the negotiation of the Treaty of 1707 unionists had doggedly pursued the goal of economic union. For both the unionist pamphleteers in 1603–5 and the commissioners of 1604 it had possessed the highest priority. Bacon argued in the draft preamble of the Instrument of 1604 that 'for so much as the principal degree to union is communion and participation of commerce . . . it appeareth to us . . . that the commerce between both nations be set open and free'.[2] Every single project for union since that time had involved some sort of commercial union. Even the union proposals of the Scottish commissioners in 1641 and 1648, which were intended mainly to promote religious union, included a provision allowing the Scots the right to transport commodities from Scotland to England and Ireland as if they were transporting them from one part of Scotland to another. In 1668 Anglo-Scottish negotiations centred exclusively on a proposed commercial treaty, while in 1689 Scotland sought a union 'in parliaments and trade'.[3] The abortive negotiations of 1702 dealt with proposals for an incorporating union which had economic terms similar to those approved in 1707. Long

[2] *L&L* iii. 243.
[3] Hughes, 'Negotiations for a Commercial Union'; Smout, 'Road to Union', 184.

before 1707, therefore, Englishmen and Scots, even if they were opposed to union, assumed that any union agreement would entail some sort of commercial accord between the two nations and at the very least a reduction of the trading barriers that separated them.

Commercial union may have been a predictable feature of the Treaty of 1707, but the acceptance by both Scotland and England of the economic terms of that Treaty calls for some explanation. On the one hand, we must explain why Scottish opposition to an economic union, which became much stronger in the early eighteenth century than it ever had been before, was not able either to prevent the conclusion of the Treaty or at least to secure more protection for Scotland's commercial 'interest' and greater economic autonomy for Scotland within the union. On the other hand we must explain why English opposition to a commercial union, which had been largely responsible for the failure of the union projects of 1604 and 1668, did not either prevent English ratification of the Treaty or at least reduce the number and extent of the apparent concessions that England made to Scotland.

In response to both questions, one can argue that the Englishmen and Scots who negotiated and voted for the Treaty desired union for mainly political and other non-economic reasons and were therefore willing to accept an economic union that was less than optimal from their country's point of view. One can also argue that the terms of the Treaty represented a compromise in which the Scots accepted higher taxation and British regulation of their economy in exchange for participation in English trade and the receipt of so many concessions. But even if unionists subordinated economic to political considerations and were willing to make commercial and economic compromises, they were unlikely to have agreed to a Treaty that they thought would have harmed their country's economic interests. In other words, the Treaty of Union would never have been successfully concluded if the arguments against economic union had been compelling. This chapter will first explore the arguments that unionists advanced in favour of economic union. It will then study Scottish arguments against such a union in order to explain why those arguments failed to have a significant impact on the men who negotiated and eventually voted to accept the Treaty. Finally, it will study English arguments against an economic union and will show why these arguments, which had a demonstrable effect on union negotiations in the early seventeenth century, lost most of their persuasive force in the early eighteenth century.

I

Much of the support for an economic union of England and Scotland stemmed from the belief that free trade and an equality of customs went hand in hand with a union of the kingdoms. Since both Scotland and England possessed internal free trade, it seemed only natural that a united Britain would have the same freedom. This axiomatic association of commercial union on the one hand and a union of the kingdoms on the other is one reason why all formal projects for union between 1603 and 1707 included a provision for some sort of economic union. Advocates of an incorporating union were not the only ones who made this association. Those who wished to bring the two kingdoms closer together without uniting the parliaments—those who simply wished to promote a 'union of love' between the two nations—seemed just as eager as the advocates of full incorporation to effect an economic union. The commissioners of 1604, for example, who did not even consider a union of parliaments, proposed the elimination of most of the trading barriers between the two countries. Likewise the Scottish commissioners of 1641 and 1648, who certainly had no intention whatsoever of uniting the parliaments, argued strongly for freedom of trade. For them the freedom of commerce was to be symbolic of the closer relationship between the two peoples that they wished to foster. From their point of view there could be 'no greater mark of mutual amity betwixt the kingdoms than a fair and peaceable conversing at home and abroad'.[4]

Although unionists considered economic union to be a natural corollary of a closer union of the kingdoms, they frequently discussed it on its own terms and advanced strictly

[4] Harl. MS 455, f. 22ᵛ.

economic arguments in its support. Very often they stressed the economic benefits that both countries would receive if such a plan were to be implemented. They argued that a commercial union would increase the number of commodities that would become available to both nations at low, customs-free prices. Even if one or two trades in either country might suffer from the competition of their new countrymen, the general good of all Britain would be served. Union also promised to give to a united Britain a larger share of the world's trade than either England or Scotland could possess by itself, much in the same way that it promised to increase the military power and diplomatic prestige of both countries. The old adage that in union there was strength seemed to have a useful economic interpretation here, especially for those who thought in mercantilist terms of gaining the largest possible share of world trade.

When unionists sought to win support in one country or the other, as they so often did in the early eighteenth century, they spoke less of mutual advantage and emphasized the benefits that their audience could expect to receive from the union. Their task was for the most part easier in Scotland than in England for the simple reason that throughout this period Scotland was the poorer and less economically developed of the two countries and therefore stood to gain by associating itself with a wealthier partner. Paramount among the benefits that the unionists claimed Scotland would receive was admission to the English colonial trade. As is well known, Scotland had a singularly unsuccessful record in its colonial ventures, the two great failures being the Nova Scotia project in the early seventeenth century and the Darien scheme in the 1690s.[5] Not only did it have no colonies of its own, but after 1660 it was excluded by the English Navigation Act from participating in trade with English colonies. A communication of trade between the two countries, a central feature of economic union, would make Scotland a joint custodian of an established and growing empire and allow it to trade with all English colonies on an unrestricted basis.

Participation in English trade also promised to give

[5] G. P. Insh, *Scottish Colonial Schemes* (Glasgow, 1922).

Scotland, as part of a united Britain, a share of a highly profit-
able overseas trade throughout the world and remedy its own
unfavourable balance of trade, which in the 1690s had become
acute.[6] In this way Scotland would also be able to reduce its
dependence on trade with England, which in the seventeenth
century had become its primary market. At the same time,
however, Scotland would profit from the elimination of
customs duties on trade with England. Its merchants would
be able to sell cloth, cattle, salt, and grain at duty-free prices,
while others would pay less for English commodities they
desired, such as wool for the cloth industry. Finally, the union
promised to attract English capital and skill to Scotland and
thereby serve the general purposes of economic development.
Sir Thomas Craig recognized this last potential benefit of
union as early as 1605, and his argument won considerable
support in the pro-union pamphlets of the early eighteenth
century.[7]

In enumerating the benefits that England could expect to
derive from economic union, pro-unionists were faced with a
more difficult task. Defoe, to be sure, argued that the English
would profit from the union more than the Scots, since they
would be able to sell their goods (especially wool) at lower
prices in Scotland and thereby undercut their Scottish
competitors.[8] Defoe, however, could not deny the fact that
Scottish trade was considerably less important to England
than English trade was to Scotland, and this reality weakened
any argument based on comparative advantage.

Somewhat more compelling was the argument that England
would gain access to the abundant fisheries off the coast of
Scotland. Here at least was one area of economic activity in
which the English clearly stood to gain from union. In the
early seventeenth-century discussions of union the Scottish
government had tried to retain for its own fishermen the
exclusive right to use these waters. The Instrument of 1604,

---

[6] T. C. Smout, *Scottish Trade on the Eve of the Union, 1660–1707* (Edinburgh, 1963).

[7] Craig, *De Unione*, 273; Chamberlen, *Great Advantages*, 27; Buchan, *Memorial*, 6.

[8] [Defoe], *Essay at Removing National Prejudice*, Pt. I, 30–1. One Scottish writer
objected to the union precisely because he thought the English would profit in this
way and Scottish manufacturers would suffer. NLS, MS 1019, f. 2. For a similar but
vaguer view in the early seventeenth century see [Cornwallis], *Miraculous and Happie
Union*, sig. C4ᵛ.

for example, contained a clause reserving to each nation the
right to the surrounding seas for a distance of fourteen miles.
In 1631 Charles I attempted to open all of these waters to a
joint Anglo-Scottish fishery company as part of his efforts to
challenge Dutch fishing supremacy in the North Sea. The
Scottish commissioners who discussed the plan with English
commissioners in London tried to reserve their traditional
rights within the fourteen-mile limit, but the strength of
English opinion in favour of the project, the king's persistence,
and the constitutional fact that the seas, coming within the
ambit of the king's prerogative, were already united by the
Union of the Crowns forced the Scots to yield.[9] As it
was, the project failed miserably, but the English continued in
the late seventeenth century to eye the fisheries jealously. In
addition to giving the English an opportunity to increase their
export of fish to the Continent and thus to challenge the
Dutch, access to the fisheries promised to give England a
'nursery of seamen' so that the country could adequately
maintain its navy.[10] An economic union of the two countries
would make this possible. On this point the English gained
some support from Scottish unionists. Since the Scots failed to
develop their own North Sea fishing industry in the late seventeenth century, writers such as John Buchan and the Earl of
Cromarty argued persuasively that a joint British effort was
necessary in order to overcome the Dutch.[11] Both countries,
according to William Seton, stood to benefit from such an
enterprise.[12]

In addition to fish from the North Sea, unionists claimed
that England would, as a result of the union, gain access to
many more Scottish commodities than it already imported.
They had difficulty, however, naming products that the
English urgently required. Certainly Cromarty was vague
when he referred to the 'inexhaustible mines of trading stuff'

[9] T. W. Fulton, *The Sovereignty of the Seas* (Edinburgh, 1911), 222–36; Lee, *Road to Revolution*, 102–5.

[10] Sir W. Monson, 'A Discovery of the Hollanders' Trades . . .', in J. Churchill, *A Collection of Voyages and Travels* (London, 1732), iii. 465–500; [Sir W. Petyt], 'Britannia Languens', in *Early English Tracts on Commerce*, ed. J. P. McCulloch (Cambridge, 1952), 303.

[11] Buchan, *Memorial*; Mackenzie, *Parainesis Pacifica*, 5.

[12] Seton, *Interest of Scotland in Three Essays*, 59.

lying untapped and untraded in Scotland's three firths.[13] John Spreull, a Scottish pamphleteer whose position on the broader question of union was non-committal and whose main objective in writing was to establish Scotland's economic self-sufficiency, did compile a list of Scottish commodities to prove 'that tho England join with us in union or communication of trade, they will not be married to a beggar, with whom they should find nothing but a louse in our bosom the first night'.[14] Although this list was long, including twenty-four different types of goods, it consisted of little more than fish (which the English admittedly did desire), cattle (which they already imported), and a host of mineral products, such as coal, lead, marble, silver, and gold.[15] The demand for most of these mineral products, however, was not great, and early seventeenth-century discoveries of Scottish gold and silver, which had attracted much notice in England, had not resulted in large enough yields to have served as a persuasive argument in favour of free trade.[16]

In many ways the most attractive item in Spreull's list of Scotland's 'products and manufactures' was not a material but a human commodity: a hundred thousand fighting men to serve by land and sea. Since England sought union in the early eighteenth century mainly for purposes of national security, the acquisition of military power in this form had a high premium upon it. The addition of the Scots to the English population also appealed to English unionists on more strictly economic grounds. When Defoe referred to the 'inexhaustible treasure of men' that the union would bring to England, he had in mind not merely Scottish seamen and soldiers but Scots who could staff England's industries and inhabit her colonies.[17] In the early eighteenth century many English economic thinkers had begun to measure the wealth

---

[13] G. Mackenzie, Earl of Cromarty, *Two Letters Concerning the Present Union, from a Peer in Scotland to a Peer in England* (Edinburgh, 1706), 10.

[14] J. Spreull, *An Accompt Current betwixt Scotland and England Ballanced* (Edinburgh, 1705), dedication. For a discussion of Spreull see Smout, 'The Road to Union', 186.

[15] 'A Scheme of Scotland's Product and Manufactures', appended to Spreull, *Accompt.*

[16] *Illustrations of British History*, ed. E. Lodge (3 vols., London, 1791), iii. 224; Winwood, *Memorials of Affairs of State*, ii. 422–3.

[17] Defoe, *Essay at Removing National Prejudice*, Pt. I, 27–8, II, 30.

of a country by the size of its population, especially its employed population, rather than by its volume of trade. Sir Francis Brewster wrote that 'ordinary people ... being more in number do more contribute to increase the nation's wealth', while Peter Paxton argued that the number of people in a country and its wealth stood in direct proportion to each other.[18] William Petty actually assigned a value of £80 to each new addition to the English population. The arguments of this new school of 'social mercantilism' conveniently served the cause of union. Hugh Chamberlen, for example, used Petty's figures to show that England, by embracing the Scots in a united Britain, would thereby become so much the wealthier.[19]

Even if Englishmen could not be convinced that the union would offer them new markets, new commodities, new investment opportunities, or additional manpower, they still might recognize that it would offer them better protection of the economic assets that they already had. As Chamberlen argued in response to the assertion that England had little to gain from the union, the Treaty of 1707 would enable them better 'to defend and protect the wealth and trade they have'.[20] From the English point of view the most attractive feature of economic union was the *control* that a British parliament dominated by Englishmen would acquire over Scottish economic activity, both in Scotland and England. The need for such control had been apparent in England ever since the Union of the Crowns had led to an increase in Anglo-Scottish trade.[21] When that trade had been relatively free, as in the early seventeenth century, English MPs and officials had become keenly aware of the opportunities that Scots had to abuse the system, mainly by re-exporting duty-free English goods abroad to France, where Scotland had favourable trading advantages. The Instrument of Union had tried to prevent such abuses, but even if it had been approved, it

---

[18] C. Wilson, *England's Apprenticeship 1603–1763* (London, 1965), 354.

[19] Chamberlen, *Great Advantages*, 9, 25.

[20] Ibid. 13.

[21] For figures concerning Anglo-Scottish trade see A. M. Millard, 'The Import Trade of London, 1600–1640', Ph.D. thesis (London, 1956), App. C; Lythe, 'Union of the Crowns', 226.

would not have given the English the authority to enforce such restrictions themselves. Enforcement would have remained in the hands of the Scottish privy council and parliament.[22] One of the main attractions of Sandys's proposal for perfect union as an alternative to the Instrument was precisely that it would have given the English government control over Scottish economic activity. A century later Queen Anne, in proposing union negotiations, echoed Sandys's argument when she mentioned as one of the benefits of union the effective prohibition of the export of wool from Britain.[23]

II

However attractive the idea of economic union appeared to be, both to Englishmen and Scots, the prospect aroused considerable opposition in both countries. In discussing this opposition, it is necessary not only to distinguish between English and Scottish sentiment but also to show how opinion in both countries changed over time. In Scotland opposition to economic union was relatively weak in the seventeenth century but increased significantly in the early eighteenth century.[24] In England, by contrast, opposition remained strong throughout the seventeenth century but almost disappeared by 1707. These fluctuations of economic opinion reflected changes in the proposed terms of union as well as in the ways that the subjects of each country viewed the value of trade with the other.

It took a few years after the Union of the Crowns for a clearly identifiable body of opinion against economic union to develop in Scotland. From 1603 to 1606 there appeared to be nearly unanimous Scottish support for establishing at least a limited communication of trade between the two countries. The Scottish council took the first step in this direction shortly after James's departure for England by lifting the inward and outward customs on trade with England, more than a year before the English stopped collecting from the Scots the extra

---

[22] For the control of the Scottish economy by the privy council see *RPCS* vii. 304.

[23] Chamberlen, *Great Advantages*, 26.

[24] A. M. Carstairs, 'Some Economic Aspects of the Union of Parliaments', *Scottish Journal of Political Economy*, 2 (1955), 64–72.

twenty-five per cent 'petty custom' collected on aliens importing goods to England.[25] When the commissioners for the
Union agreed upon the economic clauses of the Instrument of
1604, by which a broadly conceived commercial reciprocity
between the two countries was to have been established, both
the Scottish merchants (who answered the objections of the
English merchants to these proposals) and the Scottish parliament, which voted to accept the Instrument, supported the
plan.[26] Since the Instrument did not subject Scottish trade to
English regulation, deny the royal burghs their monopoly of
foreign trade, or deprive the Scots of their exclusive fishing
rights in Scottish waters, the strength of Scottish opinion in
favour of the Instrument should not surprise us. Nevertheless, some Scots soon began to have second thoughts about the
desirability of free trade with England. The customs farmers
complained about their loss of revenue stemming from the
suspension of the Scottish customs, and in 1611 the council
decided to reimpose them.[27] In so doing they were taking
action not only to restore their lost revenue but to prevent
English producers from underselling their Scottish
competitors (who had complained to the council) and to keep
Scottish money in the country.

The reimposition of the Scottish customs did not mean
that Scotland had abandoned its hope for some sort of
commercial treaty or even union with England; it meant only
that in pursuing this goal it would not make concessions to
England that threatened to harm its economy. Since England
had never fully reciprocated by lifting all of its customs levied
on Scottish imports, the reimposition of the customs made a
good deal of sense from the Scottish point of view. The same
reluctance to grant unilateral concessions to England
manifested itself in 1631, when Scottish commissioners, as
mentioned above, resisted the efforts of Charles I to open the
North Sea fisheries to joint English and Scottish activity. In

[25] Lythe, 'The Union of the Crowns', 221–2; *RPCS* vii. 80, 347, 577.
[26] SP 14/24/5; *APS* iv. 366–71.
[27] *RPCS* vii. 392; Lee, *Government by Pen*, 84–6. The customs were enforced only
'spasmodically', however, until the time of the Bishops' War. T. Keith, 'The
Economic Condition of Scotland under the Commonwealth and Protectorate', *SHR* 5
(1908), 274.

taking this position, it should be emphasized, the Scots were not completely turning against the idea of economic integration with England. In 1604 they had taken the same position while supporting an almost complete communication of trading privileges and the elimination of the customs. Scottish opposition to freedom of trade with England was never unqualified in the seventeenth century. There were too many advantages to be gained from the elimination of the customs and from participation in English trading routes for full-scale opposition to develop. What appeared to be opposition was in fact an attempt to minimize concessions to England and maximize English concessions to Scotland. Until the late seventeenth and eighteenth centuries the idea of making Scotland completely independent of England did not attract any identifiable support in Scotland.

The weakness and limited nature of Scottish opposition to economic union with England became obvious in the middle decades of the seventeenth century. The reimposition of a number of English restrictions on trade with Scotland during the Bishops' War obviously proved to be detrimental to Scotland. Scotland had never achieved complete freedom of trade with England in the early seventeenth century, but the repeal of many of the hostile laws prohibiting trade in certain commodities, the lifting of the alien customs, the irregular collection of the ordinary customs, and the generally closer ties between the two countries since 1603 had resulted in a dramatic increase in Scottish trade with England. In order to restore this trade, as well as to encourage the further union of the two peoples that the Covenanters desired, the Scottish commissioners in 1641 proposed a complete freedom of commercial intercourse between the two nations. They also sought permission to trade freely in any place where the English had an 'out-trade', and they asked that Scots be given all the 'benefits and privileges of natural born subjects'.[28] This was a plan for economic union to which few Scots could reasonably object, and there is no evidence that any did in fact oppose it. If enacted, the proposals would have given the Scots a number of concessions without asking very much from them

[28] Harl. MS 455, ff. 21–23.

in return. For this reason the proposals encountered strong
opposition from the English commissioners.[29]

The only time before 1690 that Scottish opinion against
economic union appears to have gathered real strength was
during the 1650s, when as part of the Cromwellian Union
complete freedom of trade between the two countries was
established. It was of course difficult to dissociate the
economic union of the 1650s from the broader political union
with which it was connected, but there is some evidence that
the commercial union, considered by itself, was unpopular in
Scotland.[30] Since the terms of this union differed considerably
from those sought by the Scottish commissioners in 1604 and
1641, the emergence of negative Scottish opinion regarding
the scheme is not surprising. As it is, however, we cannot be
certain that Scots uniformly or unequivocally opposed the
new arrangements. The Scottish economy, while hardly thriv-
ing in the wake of a civil war and an invasion, clearly
benefited from the English connection, and some Scottish
trades, especially salt and linen, prospered. It was mainly for
economic reasons that the end of the Cromwellian Union was
viewed by some Scots, and continues to be viewed by many
historians, as a mixed blessing.[31]

In any event, the end of the Cromwellian Union and the
development of a fierce economic rivalry between England
and Scotland after the Restoration put the question of
economic union on an entirely new basis. After 1660 economic
union meant the reduction or elimination of the extraordinarily high protective duties that England and Scotland imposed
on imports from the other country and the relaxation of the
English Navigation Act so that Scottish ships would not, as
the Act provided, be considered as if they were foreign vessels.
It appears that Scots generally favoured such a union. It may
be true that Scottish merchants did not suffer as much as they
claimed from the English navigation system, but they clearly
desired its modification in favour of Scotland. It may also be
true that the Scots 'did not desire free trade', as Riley has

[29] Hamilton, 'Anglo-Scottish Negotiations', 85.
[30] Keith, *Commercial Relations*, ch. 3.
[31] Omond, *Early History*, 118–21; Ferguson, *Scotland's Relations with England*, 137–41.

claimed,[32] in that they may very well have wished to retain import duties on certain English goods, but they clearly resented high English duties and hoped, as in the past, to secure a reduction of them.[33] When Scottish commissioners, therefore, tried unsuccessfully in 1668 to negotiate a commercial treaty with the English, according to which a limited freedom of trade would have been established and the Navigation Act relaxed for Scotland, they were almost certainly representing the interests and the desires of their countrymen. As in the past, Scots may have had certain reservations about economic union, but as long as the pursuit of economic union meant winning concessions from the English, it was difficult to take a strongly negative position.

It was not until the early eighteenth century that a strong and cohesive body of opinion against economic union emerged in Scotland. This opposition, which had only a weak precedent in the 1650s, arose in response to changes in the terms of the proposed union and a new analysis of Scotland's economic condition. The change in the terms of union was that, beginning in 1670, all union negotiations centred on proposals for an incorporating or parliamentary union. Proposals for economic union continued to be discussed, as before, but now within the context of a broader plan for political union. The prospect of parliamentary union meant that from the Scottish point of view an economic union would be qualitatively different from what they had considered and debated during the early seventeenth century. Instead of the reduction of trading barriers by the joint action of two separate parliaments which would retain control of the commercial and economic life of their respective countries, they were now dealing with the proposed assumption of the economic control of their country by a parliament dominated by Englishmen. In the same way that a united British parliament threatened their law and their church, it also threatened their economic independence.

The second reason for the increasingly strong Scottish

---

[32] Riley, *Union*, 202, 245 n. Hodges, *Rights and Interests, Treatise I*, 10, argued that free trade would bring little or no benefit to Scotland, 'at least without limitation'. See also the reservations expressed in Seafield's 'Memorial'. HMC, *Laing MSS*, ii. 132–3.

[33] Hence Lauderdale appealed to the Scottish parliament in 1669 for a union mainly on these terms. NLS, MS 597, f. 215. See also the appeal of James VII in 1686. *APS* viii. 579.

opposition to economic union was the growing belief that the Scottish economy, which was in terrible shape in the 1690s and early 1700s, had become depressed mainly because of the 'English connection'.[34] Whether or not this perception represented a valid assessment of the cause of Scotland's economic woes is not all that important. It probably did not. The Union of the Crowns had caused an unhealthy dependence of the Scottish economy upon the English market,[35] but the main causes of Scotland's economic problems were war, famine, and the failure of the Darien scheme, not English protectionism.[36] What is important is that a large number of Scots attributed Scotland's economic malaise to the increase in English influence since the Union of the Crowns. Although some unionists used this analysis of Scotland's economic predicament as an argument in favour of a union, claiming that only complete economic integration would end the anomalous relationship that prevailed during the regal union, anti-unionists feared that further union would only place Scotland in a more dependent and subordinate position.[37] Just as the Cromwellian Union had brought about the encouragement of those Scottish industries that catered for an English market, so the proposed union of 1707 would subordinate Scottish to English interests.[38] As one Scottish pamphleteer predicted, the Scots would be made 'footstools of England'.[39]

The prospect of parliamentary union, therefore, coupled with the recognition that the English connection lay at the root of Scottish economic difficulties, accounts for the dramatic increase in the strength of Scottish opposition to economic union at the beginning of the eighteenth century. For the first time in one hundred years there existed a substantial body of Scottish opinion against the proposed economic integration of the two countries. This opinion did not, however, lead to any meaningful concessions to Scottish economic indepen-

[34] See for example [Ridpath], *The Reducing of Scotland by Arms*, 7.

[35] T. C. Smout, 'Scotland in the Seventeenth Century: A Satellite Economy?', in *The Satellite State* ed. Dyrvik *et al.*, 14.

[36] Riley, *Union*, 197, 201.

[37] Fletcher, *State of the Controversy*, 14–15.

[38] *Records of the Convention of the Royal Burghs of Scotland*, iv. 400–1, cited in Carstairs, 'Economic Aspects', 66–7; Riley, *Union*, 230–1.

[39] *The Smoaking Flax Unquenchable*, 8.

dence in the negotiation of the Union Treaty. Nor did it significantly influence the members of the Scottish parliament when they voted to ratify that Treaty. Article 4 of the Treaty, which established a communication of trade and navigation, passed the Scottish parliament with astonishing ease, receiving only nineteen negative votes.[40] It has been argued that Scottish 'opinion' on the union, especially as expressed in the pamphlets of the early eighteenth century, did not have a decisive impact on the outcome of the union question, since the English government had already secured the necessary votes by various and often devious political means.[41] That may well have been the case, but Scottish opinion had at least indirectly influenced the shape of the Union Treaty, especially with respect to law and religion, and it accounted for many of the much larger negative votes that were recorded on other articles of the Treaty. The question remains, therefore, why Scottish opposition to economic union had such a limited political impact.

Part of the answer lies in the ambivalence of much of this negative opinion. Most of the Scots who argued against the Union Treaty on economic grounds did not oppose all forms of economic union. William Black, for example, who was by far the most prolific of the economic anti-unionists, could never bring himself to support a plan for complete economic separation. Black argued as cogently as any pamphleteer that Scotland did not need an economic union with England. He claimed that Scotland's unfavourable balance of trade (with all countries) had lately shown signs of improvement and could even become favourable if the government would more strictly regulate the purchase of foreign-produced luxury goods. Black also threw cold water on the exaggerated claims that unionists had made regarding the possibility of a post-union Scottish economic miracle, and he criticized many of the particular features of the Union Treaty, such as the provision for Scotland paying taxes at the English rate. Nevertheless, Black favoured a limited communication of trade, which he hoped could be negotiated within the context of a

---

[40] Ferguson, *Scotland's Relations with England*, 261.
[41] Riley, *Union*, 245.

federal union. In the final analysis, Black was far more opposed to an economic union that was part and parcel of an incorporating union than with the prospect of economic union as such.[42]

The patriotic speeches and writings of Andrew Fletcher and James Hodges, two of the most outspoken critics of an incorporating union, were no less ambivalent than those of Black on the question of economic union. Fletcher left the Scottish parliament in anger when he saw that Article 4 was passing by a large margin,[43] but he himself had envisaged some sort of communication of trade as part of the federal union that he advocated. He regretted as much as the unionists themselves that the two countries had not agreed upon a communication of trade in the early seventeenth century and in 1689 he himself had favoured a union of trade.[44] In a similar vein James Hodges listed as one of the mutual advantages of a 'happy union', which from his point of view could only be federal in nature, the 'mutual benefits which each may gain by freedom and communication of trade'.[45] To be sure, Hodges did insist upon articles of limitation to preserve the trade of Scotland, but his support for the basic principle of commercial union weakened the force of his specific arguments against the type of union proposed in 1707.[46]

Arguments against an economic union failed to win widespread political support not only because they were ambivalent but also because they were defensive, weak, and unconstructive. Most of the pamphlets written against economic union in the early eighteenth century represented responses to English unionist propaganda. Since most Englishmen believed that Scotland would profit most from the economic terms of the union, English unionist pamphleteers emphasized economic themes in their attempt to win support for the Treaty. The writings of men like Black had as their main objective the rebuttal of the exaggerated claims of union

---

[42] W. Black, *Essays upon Industry and Trade* (Edinburgh, 1706); *A Letter Concerning the Remarks upon the Considerations of Trade* [n.p., 1706]; *A Short View of our Present Trade and Taxes* [Edinburgh, 1706]; *Some Considerations in Relation to Trade* (Edinburgh, 1706).

[43] Carstairs, 'Some Economic Aspects', 68.

[44] Fletcher, *State of the Controversy*, 14, 25.

[45] Hodges, *Rights and Interests, Treatise I*, 15.

[46] Ibid. 26, 61.

propagandists. Although the Scottish writers succeeded to some extent in this undertaking, their arguments appeared to be defensive and to lack initiative.

The only way that Scottish anti-unionists might have changed this defensive posture would have been to devise a constructive, practicable alternative to complete economic integration. This they failed to do. Although many of them boldly asserted that Scotland could prosper on its own, they never formulated a plan for Scottish economic recovery and growth in which the English connection would be completely severed.[47] They spoke of greater restraint in the consumption of luxury goods, but even they must have realized that economic recovery required more than that. The manifest impossibility of completely independent recovery probably explains why, in the final analysis, the anti-unionists favoured a limited communication of trade within a federal union rather than complete separation. This plan, however, proved to be no more practicable or realistic than a severance of the English connection, for it amounted to little more than unilateral English concessions to Scottish trade, which the English had no reason to grant.

Scottish arguments against economic union were therefore ambivalent, defensive, and unconstructive, and they manifested far less coherence and consistency than those of their unionist adversaries. They were unlikely, therefore, to have won the broad-based political support in Scotland that was necessary to bring about a defeat or a modification of the Union Treaty. Aside from their intrinsic weaknesses, these arguments failed to influence the course of events because they were formulated belatedly. The Scottish case against economic union did not, like the corresponding cases against legal and religious union, have a long and respected pedigree. It had not served as a standard demand of Scottish commissioners for the union throughout the seventeenth century and it had not, therefore, won the recognition of English unionists as a Scottish interest that required special consideration or protection. By the time serious and vigorous Scottish opposition to economic union had emerged, the terms of the union

---

[47] Smout, *Scottish Trade*, 269–70; Riley, *Union*, 214, 231.

and the extent of English concessions to Scottish independence had already been determined.

The case for Scottish economic independence within the
union also suffered because it lacked a distinct institutional
focus. Opponents of legal and religious union had separate,
well-established national institutions, the law courts and the
church, whose preservation within the union they could
reasonably demand. Opponents of economic union had no
corresponding institution to serve as a focal point of their
demands. Scottish trade was regulated by parliament and the
privy council, the very institutions whose abolition an incorporating union entailed. This meant, in practical terms, that
the question of economic union could not easily be dissociated
from that of political union. The only way to protect Scottish
commercial interests in a united Britain would have been to
place a series of limitations on British parliamentary power or
negotiate a federal union. From the English point of view,
neither of these options was acceptable.

When all was said and done, the Scottish case against
economic union was not especially strong, and therefore it had
little hope of bringing about either a defeat of the Union
Treaty or a modification of its terms. One of the clearest signs
of the weakness of this anti-unionist position was the relatively
limited attention that Scottish opponents of the Treaty gave to
economic questions. Very few of their pamphlets and speeches
in the early eighteenth century dealt exclusively with
economic issues, while those which discussed the full scope of
the union question placed very limited emphasis upon its
economic aspects. To some extent this de-emphasis
represented a strategic consideration. Since English unionist
propaganda had used the prospect of Scottish prosperity to
win support for the Treaty, and since the anti-unionist
arguments were weak and defensive, the anti-unionists sought
to soft-pedal the entire question and to focus on the broader
and more substantial questions of sovereignty and religion
that the Treaty raised.[48] In so doing, however, the anti-
unionists also made a statement regarding their own
priorities. For them the preservation of their parliament, their

---

[48] Riley, *Union*, 219.

separate administration, their church, their law, and even the rights of the burghs and their heritable jurisdictions meant much more than the preservation of their economic independence. In short they were concerned with their traditional 'liberties', among which the Scottish commercial interest could not easily be included.

III

English opposition to commercial union differed considerably from the corresponding Scottish sentiment. Instead of reaching its zenith in the early eighteenth century, it achieved its greatest strength in the period before 1670 and became noticeably weaker after 1702. During the early seventeenth century it manifested none of the ambivalence that later characterized Scottish anti-unionist thought, and it possessed as much, if not more strength than English opposition to political, legal, or religious union. English thought on commercial union was, moreover, more blatantly xenophobic than its Scottish counterpart, since it was assumed that Scots would come to England to pursue their mercantile interests rather than vice-versa. Finally, English arguments against commercial union had much more cogency than their later Scottish counterparts.

In describing English opposition to economic union, it is appropriate to look first at the expression of such sentiment during the period 1604–7. During those years English opponents of union attached the greatest importance to economic questions and argued passionately against it on those grounds. In so doing they anticipated nearly all the arguments against a commercial union that would appear during the following one hundred years. It must be remembered, however, that Englishmen in the early seventeenth century argued against a commercial union that was to be established without a corresponding parliamentary union. Many of their arguments would have lost their force if James had succeeded in realizing his original goal of uniting the two parliaments.

Englishmen advanced three main arguments against a commercial union. The first and by far the most persuasive was that if such a union were to be negotiated, Scotland would gain an unfair competitive advantage with the English in overseas trade and thus deprive English mariners and merchants of their livelihood. The clearest illustration of the way in which the Scots would profit at English expense was in the French trade. Since Scots became naturalized in France at birth and could purchase letters of burgess in Normandy and Bordeaux for a mere 7s., they did not have to pay the same customs as English merchants trading there.[49] It appeared, therefore, that if Scottish merchants were to be given freedom of trade in England, they would be able to undersell their English rivals either by exporting duty-free French goods into England or by exporting English goods to France.

A careful study of Scottish trade with France, undertaken in December 1604, revealed that the Scots would have to practise deception in order to gain the competitive edge that the English feared. The exemption of goods transported from France was granted only if the goods were intended for a Scottish market, and a certificate from Scotland was required to confirm that Scotland was indeed their ultimate destination. If the goods were intended for re-exportation to England, the Scottish merchant was required to pay the full French duty as well as the English inward customs. Re-exportation from Scotland was at least theoretically even more costly, because both inward and outward Scottish customs would also be collected. The English merchants feared, however, that Scottish ships, possessing a certificate, would none the less head for England or, after landing in Scotland, re-export the goods to England either by ship or across the Border. The possibility certainly existed, but in order to meet the English objection, the Scots agreed to pay additional customs in England to offset the advantages they would gain in France. The problem of overland re-exportation could have been resolved by ensuring that customs be paid at Berwick or Carlisle, while a provision such as the one included in the Instrument of Union against exporting English wool to

---

[49] SP 14/24/11. An Englishman could also become naturalized in France, but he had to pay 100 crowns (£25). In order to become a burgess he had to take an oath of allegiance and thus abdicate his English citizenship. See Keith, *Commercial Relations*, 13–14.

Scotland might have succeeded in solving the converse problem of re-exporting English goods to France. To appease the English merchants on this score, Scottish merchants also offered to carry English cloth to France under the name of Scotland, so that it might be freed of French customs.[50]

Even if the sticky problem of the French trade could have been resolved, the English still feared that the Scots would profit from a commercial union in other ways. A main source of concern appears to have been the prospect of Scottish ships stealing a large portion of the English carrying trade. The Instrument of Union specifically provided for the shipping of commodities in English or Scottish bottoms 'indifferently', and although the Instrument was never ratified, James did eventually designate Scottish ships as 'free bottoms' in England by proclamation in 1615.[51] To such a policy most English merchants were strenuously opposed. In 1607, when the English parliament considered the Instrument of Union, the merchants of Trinity House submitted a number of articles objecting to the 'employment of Scottish ships in England'. Aside from making the general argument that England could not accommodate an increase in the number of ships trading in the country, the merchants argued that Scottish shipowners had fewer expenses than their English competitors and could therefore undercut them. Scottish owners had access to cheaper materials and labour in maintaining their ships, while their mariners, who often received their wages in the form of duty-free Scottish goods (such as knit caps and hose), did not receive the same pay as English seamen.[52] MPs in the parliament of 1607 expanded upon this point, observing that the Scottish mariner's diet was much leaner than that of his English counterparts.[53] Scottish trading practices further reduced Scots shipowners' costs. Marketing inferior goods, engaging in unethical trading practices, and avoiding the costs of landing their ordnance in France all enabled the Scottish owner to charge less to carry his merchandise.

[50] SP 14/24/11.
[51] *CJ* i. 322. Keith, *Commercial Relations*, 17–18.
[52] Harl. MS 158, ff. 165–166. See also SP 14/10A/17, f. 2.
[53] Harl. MS 1314, f. 35.

Even the relative size and character of the ships of the two
nations figured in this analysis. English ships, being large
and designed for war as well as trade, required more sailors
and had proportionately less room for storage than Scottish
vessels. The Instrument had required Scots to build larger
ships, but the English merchants claimed that Scottish
harbours, being 'dry', could not accommodate them, while the
Scots themselves confessed that there was little advantage in
constructing larger ones. If the Scots continued to build
smaller ships, and if Scottish bottoms were to gain their
freedom in England, then the results would be disastrous for
England.[54] Sir Henry Spelman predicted as early as 1604 that
'their ships shall fly with sails at sea, whilst ours lie unrigged
at the harbour', yet another example of the way in which
Englishmen feared that the Scots would gain superiority over
the English as the result of the union.[55] In the long run,
English merchants claimed, the security of England would be
endangered, since the number of English mariners would
decline, the king's ships would not be manned, and mariners
would eventually find service in foreign lands.[56]

The English merchants seemed to have an unlimited supply
of arguments to prove that the Scots would gain an unfair
advantage if a commercial union were to be negotiated. They
mentioned that Scottish merchants, not being residents of
English towns, would avoid both national and local taxes
while possessing the same opportunities to engage in trade.
They claimed that Scotland's geographical proximity to the
northern ports of Europe would allow them to undersell
southern English merchants in the overseas cloth trade. They
complained that even if the re-exportation of English goods
from Scotland would be forbidden by law, it would be difficult
to prevent the sale of such goods to foreigners in Scotland for
the purposes of exportation. Finally, the English merchants
complained that the Scots were willing to trade inferior
products in foreign markets, whereas their honour and pride
prevented them from following suit.[57] Both the Scottish

[54] Harl. MS 158, f. 166.
[55] Spelman, 'Of the Union' in *Jacobean Union*, 177. See also SP 14/10A/17, f. 3.
[56] Harl. MS 158, ff. 165–166, art. 7.
[57] SP 14/24/3.

merchants and the Earl of Salisbury responded effectively to these objections, but they did little to assuage the exaggerated fears of the English merchants that they would be seriously harmed by Scottish competition.[58]

The second main English argument against economic union concerned the composition of Anglo-Scottish trade. Reduced to its simplest form, it stated that Scotland had a greater need of English commodities than England did of Scottish goods and that therefore the reduction or elimination of commercial barriers between the two countries would benefit Scotland and harm England.[59] The argument was predicated on three assumptions. The first was that with the notable exceptions of cattle, herring, and salt, England did not need the commodities that Scotland had for sale in England. The infusion of many non-essential Scottish goods into the English market would, therefore, bring increased competition for English retailers. As Spelman wrote shortly after the Union of the Crowns, 'The commodities of their country, though they be many and good, yet as they are not such as we seek, but such as we sell, corn, cloth, wool, and such like. And to bring these to England were to bring wine to Bordeaux, apples to Calabria.'[60] If Scotland had been able to supply England with products she needed, such as wine or vegetable dyestuffs, Dr Lythe reminds us, the English approach to these questions would have been entirely different.[61] The second assumption, which reflects the old, pre-mercantilist policy of 'provision' or the prevention of scarcity, was that the commodities which Scotland desired were precisely those which the English desperately needed to keep in their own country, such as wool, leather, and tallow. The third assumption was that Scotland, being a poor country, did not possess enough wealth to purchase those goods that England wished to export.[62] 'The way to increase wealth', wrote one Englishman as early as 1604, 'is to commerce with those that are rich, for in the company of the rich the poor can be no losers; and to

---

[58] SP 14/24/4–5.
[59] Spelman, 'Of the Union, in *Jacobean Union*, 162, 170–2.
[60] Ibid. 170.
[61] Lythe, 'Union of the Crowns', 227.
[62] SP 14/24/8.

the contrary, rich men shall wax poor if they commerce with the needy.'[63] All in all, therefore, there was little to be gained from the increase in Anglo-Scottish trade that the union would encourage. Instead of Scotland becoming a market for English products while supplying greatly needed commodities, it would drain England's resources and threaten the full employment of its people.

The third English argument against free trade with Scotland was that the king would lose revenue from the suspension of the customs, an argument that appealed as strongly to the customs farmers as to the members of the central government. Anti-unionists did not hesitate to show how serious the loss would be, predicting in 1606 that the unrestrained export of cloth and other commodities to Scotland would cost the English government more money annually than the entire customs of Scotland brought in.[64] King James took some time to appreciate the merit of this argument. Even after Spelman had boldly asserted in the words of Richard I that the crown 'could not spare' the customs,[65] even after a study of the customs receipts revealed that the crown had collected £4,385. 11s. from Scottish merchants trading with England between 1596 and 1603,[66] and even after the English parliament had refused to establish the customs union proposed by the Instrument, James continued to use his prerogative power to keep the customs on Scottish goods as low as possible. Only after 1611, when he consented to the Scottish council's reimposition of their customs, did he apparently recognize that there was some value in maintaining the old policy. After that time the only break that Scottish merchants received in trading with England was that they did not pay the twenty-five per cent extra customs levied on aliens. James had lifted this duty in December 1604 and, in light of the decision in *Calvin's Case* in 1608, it was never reimposed.[67]

The English arguments used against commercial union in 1604–7 reappeared later in the seventeenth century whenever

[63] SP 14/10A/17, f. 4.
[64] SP 14/24/8.
[65] Spelman , 'Of the Union' in *The Jacobean Union*, 173.
[66] SP 14/5/27.   [67] Lythe, 'Union of the Crowns', 226.

the topic of economic union arose. They formed the basis of the English commissioners' response to the Scottish proposals of 1641 and of the criticisms that many Englishmen had of the Cromwellian economic union. The main reason why the English parliament reimposed high customs duties on Scottish imports in 1660 was the fear that Scots would undersell them, as some had done in the 1650s, while the extension of the Navigation Act to include Scotland reflected many of the concerns expressed by English merchants in 1607. By 1660, however, an entirely new argument against freedom of trade with Scotland had arisen. By that time the balance of trade between England and Scotland, at least according to one source, stood in Scotland's favour.[68] In other words, the value of goods imported from Scotland to England exceeded the value of the goods travelling in the opposite direction. In such circumstances the reduction or elimination of the customs did not make much sense, at least according to the mercantilist theory that a country should use customs duties to discourage imports, encourage production, and increase exports in order to achieve a favourable balance of trade. English knowledge of trading statistics, imprecise though they may have been, may therefore have played an important part in the reluctance of English commissioners to agree to a commercial union, especially in 1668.

In the late seventeenth and early eighteenth centuries English arguments against commercial union began to lose their force. The main reason for this was that all the formal proposals for union involved a union of the English and Scottish parliaments. This of course was the very same reason that Scottish opposition to economic union began to increase at the same time. From the English point of view a union of parliaments promised to solve many of the problems that anti-unionists foresaw in a commercial union. The anti-unionists themselves admitted that many of the 'discommodities' that they believed would arise from freedom of trade could be prevented by 'good and politic laws'.[69] The commentaries they

---

[68] D. Woodward, 'Anglo-Scottish Trade and English Commercial Policy during the 1660s', *SHR* 56 (1977), 162–4. Until 1664 English officials laboured under the false impression that England had a favourable balance of trade with Scotland.

[69] SP 14/24/8.

had made on the Instrument of Union of 1604, for example, were replete with recommendations for legislative action. Laws, they believed, were necessary to prevent the export of certain commodities to France, to equalize the customs, to solve the problem of unequal taxation, and to regulate the shipping of the two nations.[70]

Although the separate parliaments of the two nations might have been able to solve some of the economic problems connected with the union, only a united British parliament could have provided England with the guarantee that Scotland would not profit unfairly from the union or avoid the assumption of her share of economic burdens. As one English writer observed at that time, the objections against the union on the grounds that it would hurt England 'are answered by their being subject in all things to our laws'.[71] When in fact the parliament at Westminster did achieve this type of legal control over Scotland for a short period of time in the 1650s, English arguments against commercial union lost much, but not all, of their support.

The negotiations for a commercial treaty in 1668 made it even clearer that if an Anglo-Scottish economic union were to be achieved, there would have to be a union of parliaments. The official proposal of 1668, as mentioned above, did not involve any such arrangement. It dealt exclusively with commercial relations between the two countries. The problem, however, at least from the English point of view, was that the economic policies of the two countries could not be co-ordinated. In England parliament had assumed tight control over economic policy after 1660, but in Scotland the king, acting through his ministers, was still able to grant concessions by virtue of the prerogative.[72] The only way to solve the problem was to unite the parliaments.

After 1668, therefore, all English proposals for union, and indeed all formal union projects, made provision for a union of parliaments. It is not coincidental that very few English objections to the union on economic grounds arose after this time. Englishmen would continue to oppose Scottish proposals for a federal union in which there would be a limited communica-

[70] Ibid.; SP 14/24/10; SP 14/24/3.   [71] BL, Cotton MS Titus F IV, f. 57.
[72] Hughes, 'Negotiations for a Commercial Union', 33–4.

tion of trade, for this would have allowed Scotland to secure numerous commercial benefits without assuming any new burdens. They would object to the argument of men like George Ridpath that the Scots had the right to participate freely in English trading activity by virtue of their naturalization.[73] They would even threaten, as they did in 1705, to declare Scots aliens and deny them the right to export linen, cattle, and coal to England. But they would not object to an economic union on the terms that they now proposed.

The prospect of an incorporating union also had the effect of weakening the mercantilist objection to removing the customs on the goods of a country with which one had an unfavourable balance of trade. As long as Scotland was a separate state, with its own parliament, central administration, and treasury, the mercantilist argument had some validity. But if Scotland were to be incorporated into a single British state, then the argument, which was based on the retention of the greatest amount of bullion in a country, would lose its relevance. For this reason, advocates of an incorporating union had no difficulty accommodating the prospect of free trade with Scotland to mercantilist principles. To these men the freedom of Anglo-Scottish trade would constitute an extension of the freedom of English internal trade to a larger Britain, not an anticipation of the ideas of Adam Smith. It should be noted that with the possible exception of Peter Paxton, all of the English unionists, including Defoe (who is often credited with 'precocious' ideas of free trade), were committed mercantilists.[74] The union with Scotland did not require them to modify their basic principles. All they had to do was to redefine English economic interests as British interests and then seek for a united Britain the same strong position in international trade that they had previously desired for England alone.

Once Englishmen could anticipate an incorporating union with Scotland, therefore, many of their objections to economic

[73] W. Atwood, *The Scottish Patriot Unmask'd* (London, 1705), argues that freedom of trade is a right acquired from the common law and does not, therefore, accompany naturalization.

[74] J. A. W. Gunn, 'The Civil Polity of Peter Paxton', *Past & Present*, 40 (1968), 50–2; Wilson, *England's Apprenticeship*, 354.

union evaporated. At the same time, for different reasons, other familiar arguments against commercial union lost their force. By 1700, for example, the problem of Scotland's special trading privileges with France had been resolved, since the war with France had virtually eliminated Scotland's favoured position in that country. Even the advantages that England gained from the operation of the Navigation Act no longer compelled support for continued economic separation, for the enforcement of the Act had proved to be difficult and costly. To take just one example, no fewer than thirty-one Scottish ships engaged in illegal trade with the English colonies between 1691 and 1702. The English seized most of the ships, but the difficulty of enforcing the law, coupled with the loss of trade resulting from Scottish encroachments, led some Englishmen to believe that exclusion of the Scots from English trade was not worth the cost.[75]

For all these reasons there was a general weakening of English opposition to commercial union by 1707. This did not mean that all Englishmen eagerly subscribed to the arguments of the unionists. Nor did it mean that Englishmen supported the Treaty for economic reasons. That was clearly not the case. Englishmen supported the Treaty for political and diplomatic reasons that had led them to take the initiative in proposing union in 1702 and 1705. Economic motives were strictly secondary and, in the calculations of many Englishmen, did not figure at all. Most Englishmen probably considered the economic provisions of the Treaty as the price that England had to pay for union rather than a benefit that made union worthwhile. Nevertheless, the weakening of English objections to commercial union did contribute to the success of the union, for it meant a potential source of opposition to a treaty had been killed. As late as 1702 English unionists still feared that negative sentiment on economic matters might undermine efforts to negotiate a union. Accordingly, they directed some of their propaganda at the English merchant community. By 1707, however, it appeared that they did not have to worry.

[75] Graham, *Colonists from Scotland*, 15–16.

IV

The reduction of English opposition to a commercial union, the determination of the English government to secure an incorporating union, the ambivalence and weakness of Scottish arguments against an economic union, and the inability of the Scots to protect the Scottish economic 'interest' within the union all helped to ensure that one of the main features of the seventeenth-century English state—the freedom of trade from shire to shire and the regulation of the economy by the central government—would be extended to the new British state of the eighteenth century. The Scots would retain their own national bank; they would receive a new Scottish court of Exchequer; they would be compensated for the higher taxes they would have to pay; and they would receive a pledge (that was never fully honoured) that their industries would be encouraged. But there would be no separate Scottish trade, no special customs rates for Scotland, no protected manufactured goods. Just like Wales in the early sixteenth century, Scotland would become part of a predominantly English economic system and would be subjected to the economic policy formulated at Westminster.

It is not easy for us in the twentieth century, when even federal states possess highly integrated, centrally regulated, and distinctly national economies, to appreciate the novelty of the commercial features of the Treaty in 1707. No other composite political entity in Europe—indeed, no other single state in Europe in the early modern period—achieved the type of economic integration that the British state of 1707 manifested. In no other part of Europe could a merchant transport a commodity a total of 800 miles, the distance from northern Scotland to the southern coast of England, without paying one, if not many, internal customs or fees. The union of the Spanish kingdoms, for example, had done nothing to eliminate the various customs that each kingdom charged when goods passed across its borders.[76] Even France, famous for its national cohesion and political centralization, did not constitute a single free-trade zone in the manner

[76] J. H. Elliot, *Imperial Spain, 1469–1716* (Harmondsworth, 1970), 124.

of Britain. In the formation of a nationally integrated economy Britain led the rest of Europe.

The British state that emerged as a result of the Treaty of Union was, therefore, both unitary and pluralistic: unitary in its system of political and economic control, but pluralistic in its legal and ecclesiastical features. It also was pluralistic in the additional sense of containing two separate nations. The efforts that various individuals made to unite those two nations and thus to create a British nation-state will be the subject of the next chapter.

6

National Union

SEVENTEENTH- and early eighteenth-century attempts to unite England and Scotland involved two closely related but nevertheless distinct undertakings: the creation of a British state and the building of a British nation. The identity that many states and nations have achieved, especially in the twentieth century, has led to an equation of the two terms in contemporary usage, especially in Britain and the United States. Their interchangeability is most evident in the use of the term 'nationality', which can just as readily refer to citizenship or allegiance to a state as to one's membership in a nation.[1] Despite this unfortunate conflation of terminology, 'the state' and 'the nation', in the most proper senses of the words, refer to different forms of community.[2] A state is a formal and autonomous political organization, the leaders or officers of which (i.e., the government) have the legally sanctioned ability to require obedience from the inhabitants of a large and usually contiguous territory over an extended period of time.[3] It is therefore a coercive association, one which involves the exercise of power.[4] Since the eighteenth century the state

---

[1] H. Seton-Watson, *Nations and States* (Boulder, Colo., 1977), 4; T. Smith, 'Religion and Ethnicity in America', *AHR* 83 (1978, 1155). For a discussion of the 'political/ territorial' and 'cultural/ancestral' connotations of the word 'nationality' see P. White, 'What is Nationality', *Canadian Review of Studies in Nationalism*, 12 (1985), 1–23.

[2] See generally B. Akzin, *State and Nation* (London, 1964); *The Nation-State: The Formation of Modern Politics*, ed. L. Tivey (Oxford, 1981), 1–12.

[3] For somewhat different, but compatible definitions see Seton-Watson, *Nations and States*, 1; Akzin, *State and Nation*, 8; C. Tilly, 'Reflections on the History of European State-Making', in id. (ed.), *The Formation of National States in Western Europe*, (Princeton, NJ, 1975), 70; and W. Ullmann, *Medieval Political Thought* (Harmondsworth, 1975), 17, who defines it as 'an independent, self-sufficient body of citizens which lived, so to speak, on its own substance and own laws'.

[4] G. Poggi, *The Development of the Modern State: A Sociological Introduction* (Stanford, Calif., 1978), 1, sees it as 'a complex set of arrangements for rule'.

has also been an abstraction, identical with neither the gover-
nors nor the governed, to which loyalty is owed.[5] A nation is
basically what we mean by a 'people'—a large (i.e., not local)
community possessing a variable mixture of ethnic, linguistic,
and cultural similarities, a consciousness of its own distinct
identity, and the desire to express that identity in the form of
political institutions, even if that desire is not realized.[6] These
political institutions may be part of the apparatus of the state
or the machinery of government, but the nation itself is not,
like the state, a coercive association.[7]

The identification and definition of nations at any given
time in history is usually a much more difficult process than
the identification and definition of states. Scholars may differ
on the requisite size and the degree of autonomy that a
political unit must have in order to merit classification as
a state, and the territorial boundaries of states may change,
but the formal institutional structures of states and their clear
differentiation from other forms of community make their
identification both in the past and in the present a relatively
simple operation. Identifying nations is often a much more
formidable task. Nations may have imprecise territorial
boundaries; their members possess varying degrees of national
consciousness; their composition can change through a
process of assimilation; and even the sources of national
consciousness, especially the set of cultural variables that form
the basis of national identity, can themselves change. Despite

---

[5] J. H. Shennan, *The Origins of the Modern European State, 1450–1725* (London, 1974),
9, 76; H. A. Lloyd, *The State, France and the Sixteenth Century* (London, 1983), p. xvi. For
an extended discussion of the idea of the state see K. Dyson, *The State Tradition in
Western Europe: the Study of an Idea and an Institution* (Oxford, 1980).

[6] For definitions of the nation and the difficulties of such definition see R. Rose,
'The United Kingdom as a Multi-national State' (Survey Research Centre, University of Strathclyde, Occasional Paper 6, Glasgow, 1970), 2; Seton-Watson, *Nations and
States*, 5; Tivey, *The Nation-State*, 4–8; S. E. Finer, 'Military Forces and State-Making',
in *The Formation of National States*, 88; M. Hechter, *Internal Colonialism: The Celtic Fringe
in British National Development, 1536–1966* (London, 1975), 4. For the changing connotations of 'nation' and its cognates see White, 'What is Nationality?', 10. The first use
of the word to refer to a political, rather than simply an ethnic community may have
taken place in late medieval Catalonia. See J. H. Elliott, 'The King and the Catalans,
1621–1640', *Cambridge Historical Journal*, 11 (1955), 258 n.

[7] The nation, even when viewed primarily as a political association, is considered
to be a more popular and a less coercive association than the state. White, 'What is
Nationality?', 9; Tivey, *The Nation-State*, 5.

these problems, however, nations can be identified and their boundaries defined.

One of the reasons for the frequent confusion between the nation and the state is that the two communities sometimes have the same membership and the same territorial boundaries. In these national states, as they are often called, the government exercises political power over a community which considers itself to be a distinct people and which views the state as the political expression of that national identity. The national state has become an ideal in the twentieth century, and leaders of nationalist movements as well as the rulers of many states have made prodigious efforts to realize it. Nevertheless, the boundaries of many states and nations do not coincide. In some cases the nation encompasses an area much larger than that controlled by any one state; in others the state exercises authority over more than one nation. When the latter situation prevails, we generally refer to the state as being multi-national. Multi-national states are often multi-ethnic, since ethnicity serves as a very powerful source of national identity. There are, however, some ethnically heterogeneous states, such as the United States of America, which are in no uncertain terms national states.[8]

I

On the basis of the foregoing criteria, both England and Scotland were national states at the time of the Union of the Crowns. To assign such a label to either country, however, requires a number of qualifications and reservations. First of all, throughout Europe in the early modern period, the full development of national consciousness—that sense of cohesion and identity as a people without which the nation technically does not exist—was thwarted by poor communication and the strength of local and regional consciousness.[9] Individuals cannot develop a large-scale group identity if they

[8] On this phenomenon see A. Mann, *The One and the Many: Reflections on the American Identity* (Chicago, 1979). The same can be said of Switzerland. See J. Huizinga, 'Patriotism and Nationalism in European History', in *Men and Ideas: History, the Middle Ages, the Renaissance* (New York, 1959), 126.

[9] For a definition of national consciousness see K. Deutsch, *Nationalism and Social Communication* (Cambridge, 1966), 173.

do not have available a certain amount of information about their corporate activities or if their vision is restricted to the life of the local or provincial community. For these reasons national consciousness, both in England and Scotland as well as throughout Europe, was strongest within the wealthy and literate classes.

A second reservation is that neither England nor Scotland were mono-ethnic states. Both countries included areas— Wales in the one case and the northern and western Highlands in the other—that were inhabited mainly by Celtic people. In neither case had these ethnic communities been effectively assimilated to the dominant ethnic group. The governments of both states had made efforts, as part of their programmes to tighten the control of the remote areas within their dominions, to achieve a measure of assimilation. In this regard the English efforts in Wales had been more successful than the Scottish efforts in the Highlands, but in both states there were large numbers of unassimilated Celts. These Celtic communities did not, however, seek to establish distinctive political institutions.[10]

England and Scotland were, therefore, national states in which a certain measure of ethnic diversity existed. Within the dominant ethnic groups of both states there existed numerous signs of a relatively strong national consciousness. In England this consciousness probably ran stronger and deeper than in any other country in Europe. Political and dynastic factors had much to do with this. England had two central political insitutions—the English church and parliament—and before 1603 a distinctly national monarchy, all of which revealed an unusual capactiy to attract national loyalties.

The English church, being the first national church to break from Rome, gave to all politically and ecclesiastically conscious Englishmen a strong sense that they belonged to a distinct community, separate from both Rome and the Catholic powers of Europe. In the late Elizabethan period, after the defeat of the Spanish Armada, a belief developed that England was the elect nation, a people who would play a

---

[10] See Hechter, *Internal Colonialism*, 47–78.

special role in the apocalyptic drama which was about to unfold. This belief did not begin with John Foxe, as has long been believed, but in the later work of Thomas Rogers, Anthony Marten, and Thomas Brightman, and it was most forcefully argued by John Milton and his contemporaries.[11] Although puritans became the main advocates of this idea, they did not have a monopoly on ecclesiastical nationalism. The 'Arminianism' of Archbishop Laud in the early seventeenth century was every bit as nationalistic as the puritanism of those who saw in Laud a dangerous tendency towards reunion with Rome.[12]

The second national English institution, parliament, proved no less a cynosure of national loyalties than the English church. Its elected representatives may have regarded themselves as attorneys for their constituencies, but they also recognized their role as members of the highest court and legislative body of the kingdom. The prominence of parliament in the public life of the country derived not only from the constitutional powers it possessed with respect to taxation and law-making but from the absence of subordinate or rival provincial assemblies. Unlike the representative bodies of other states, especially the Estates-General of France and the cortes of the various Spanish kingdoms, the English parliament in the late sixteenth and early seventeenth centuries did not experience a decline in status, power, or privilege, and therefore its irregular meetings continued to have the effect of binding the various shires and boroughs of the country into a nation.[13] The reality of this achievement became apparent in the mid-seventeenth century, when parliament revealed that it was able to command national loyalties that the monarchy and church could not.[14]

[11] Bauckham, *Tudor Apocalypse*, 177–80, 235; Firth, *Apocalyptic Tradition*, 106–9, 167–8, 235–6. For the older view see W. Haller, *The Elect Nation* (New York, 1963), chs. 3 and 7.

[12] Kenyon, *Stuart England*, 113.

[13] On the role of parliament in this regard see M. Judson, *The Crisis of the Constitution*, 81; R. B. Manning, *Religion and Society in Elizabethan Sussex* (Leicester, 1969), 8–9.

[14] H. G. Koenigsberger, 'Early Modern Revolutions', *Journal of Modern History*, 46, (1974), 103.

Before that troubled period in English history, and especially before 1603, the English monarchy itself was a focus of national loyalty and a source of national identity. Many of the European monarchs of the fifteenth and sixteenth centuries, as part of their policy of gaining tighter control over their dominions and strengthening the central administration of the state, encouraged and exploited a sense of national community. This they did by symbolically exalting the monarchy and emphasizing their identification with 'the kingdom', which could be conceived of as both a state and a nation. José Maravall has argued that the 'policies of the great monarchs of the late fifteenth and sixteenth centuries were based on a pre-national type of community consciousness'.[15] In many cases, of course, the exploitation of this sentiment was somewhat contrived. As J. Vincens Vives has pointed out, nowhere in the early modern Europe, with the possible exception of France, did the monarchy embody a national tradition.[16] Most monarchies at this time were supranational, and many of the policies that kings followed, especially in the making of war and peace, were undertaken in the interest of the dynasty, not the nation. Monarchs still viewed their kingdoms as patrimonies and sometimes even as foreign possessions over which they presided and on to which they imposed their will, not national communities to which they themselves belonged. Nevertheless, monarchs did encourage and benefit from the growth of national sentiment, and in some cases they were able, by using the governmental machinery at their disposal, to redefine the boundaries of nationality itself.

Although the Tudor dynasty was in some sense supranational, encompassing Wales and Ireland and maintaining an antiquated claim to France, it was primarily an English royal house, and therefore it had a relatively easy task in fostering English national sentiment. Since Ireland was ruled

[15] J. Maravall, 'The Origins of the Modern State', *Journal of World History*, 6 (1960–1), 795.

[16] J. Vincens Vives, 'The Administrative Structure of the State', in *Government and Society in Reformation Europe, 1520–1560*, ed. H. J. Cohn (London, 1971), 64. See also V. G. Kiernan, 'State and Nation in Western Europe', *Past & Present*, 31 (1965), 35; W. G. East, *The Union of Moldavia and Wallachia, 1859: An Episode in Diplomatic History* (Cambridge, 1929), 4.

as a colony of England, it did not in any way call into question the identification of the monarchy with the English nation. Wales represented a different problem, but the Welsh blood of Henry Tudor, the incorporation of Wales into the English state, and the assimilation of the Welsh landed class to an English culture not only secured the loyalty of Wales to the monarchy but also laid the foundation for a new, more broadly defined 'English' or technically 'British' nationality.[17] In creating and exploiting national sentiment the Tudors also benefited from the position they occupied in two major national institutions, the church and parliament. Especially in the former, of which the monarch was the Supreme Head, the Tudors were able to command loyalties that were based on both xenophobia and the desire for religious reform.[18]

Political, ecclesiastical, and dynastic institutions were the main sources of English national sentiment in the sixteenth century, but geographical, cultural, and economic factors also played an important part. The situation of England on an island, sharing only one boundary with another state, gave Englishmen the type of isolation from other nations that fostered a sense of belonging to a distinct community. The freedom of economic intercourse between the various shires of the country encouraged those who were engaged in trade to think at least to some extent in terms of a national market and a national economy, while the government's pursuit of a policy of economic nationalism, expressed both as a drive for economic self-efficiency and a pursuit of 'mercantilist' objectives in foreign trade, also helped to foster such sentiment.[19] More important than either geographical or economic factors, however, were cultural ones. A common although dialectically diverse English language, a distinctly national, vernacular

---

[17] T. Wilson, 'The State of England, Anno Dom. 1600', 11, refers to Wales as 'part of England'. See also *Bowyer Diary*, 185; *Diary of Burton*, iv. 133.

[18] Yates, *Astraea*, 38–47; Haller, *The Elect Nation*, 82–109, 225.

[19] L. Stone, 'State Control in Sixteenth-Century England', *Economic History Review*, 17 (1947), 103–20; I. Wallerstein, *The Modern World-System: Capitalist Agriculture and the Growth of the European World Economy in the Sixteenth Century* (New York, 1976), 102. Wallerstein sees this sentiment emerging among European merchants only in the late 17th and 18th centuries, but there is much evidence for it in Elizabethan England. On the gradual emergence of a national economy during this period see Palliser, *Age of Elizabeth*, 5.

literature, a growing awareness among the literate classes of a common English history, as well as a high degree of liturgical conformity and a common body of law made it possible for Englishmen to acquire a distinct cultural identity. Since that identity could be expressed politically, through the institutions of the state, it is valid to consider the England of 1600 as a national state.

A similar, but not quite so strong a case can be made with respect to Scotland. Comparisons between England and Scotland have shown that many of the sources of English national consciousness did not operate with equal force in the northern kingdom. The Scottish state, for example, was not as highly centralized or bureaucratically sophisticated as that of England, and since the institutions of the state were essential tools in generating national consciousness, the Scottish achievement in that regard was not as impressive. One institution in particular, the Scottish parliament, never achieved as much importance as its English counterpart, and therefore it could not perform the comparable function of attracting national loyalties. Scots law, being an amalgam of many different systems, including both English law and civil law, did not acquire a distinctly national character until the late seventeenth century, while the enforcement of the law throughout the entire country was much more problematic than in England. As discussed in Chapter 4, Scottish apocalyptic thought never spawned the idea of an elect Scottish nation; rather it placed Scotland within the context of a united British elect nation or a European Protestant movement. Owing both to a paucity of the legal record and a relative weakness of the national legal tradition, the writing of Scottish national history did not develop in the same way as it did in England. Slow economic development and weak communication between different parts of the country made a Scottish national economy more of an abstraction than a reality. Linguistic diversity was much greater in Scotland than in England, and in the Lowlands the Scottish language gradually became indistinguishable from English. In the most general terms Scottish culture remained susceptible to Continental influences, especially those of France and the Netherlands. And the divisions between Highland and

Lowland culture were so strong that it was uncertain whether the kingdom consisted of one nation or two.[20]

These reservations notwithstanding, the Scottish people in 1600 possessed a sufficient amount of national consciousness and manifested sufficient ethnic solidarity to give them unquestioned status as a nation.[21] In fact, their nation was older than England. Not only did the Scots develop a range of aristocratic institutions to sustain a national community somewhat before England or France, but the conflict with England in the late thirteenth century resulted in the Declaration of Arbroath, 'the most impressive manifesto of nationalism that medieval Europe produced'.[22] The Declaration, which was the work of the lesser nobility, articulated a national identity that was genuinely collective. Even among the mass of the population there existed a sentiment that transcended the 'proto-nationalism' that other states manifested at a later date.[23] This medieval Scottish nationalism may not have persisted in its vigour, but frequent conflict with England during the ensuing two hundred and fifty years helped to maintain a strong sense of Scottish national identity.

There were of course other sources of Scottish national consciousness. One was the royal dynasty. The Scottish monarchy was not as powerful as the English, especially in the sixteenth century, but its claim to an uninterrupted descent from prehistoric times gave it advantages as a symbol of national unity that political weakness denied it. The other great source of national sentiment was the church. Even though reformers did not develop the idea that the Scots constituted an elect nation, they did view the church as the most reformed church in Europe, and they succeeded in using the church to give an order and cohesion to Scottish society that the state was incapable of providing. Only in the

---

[20] Ferguson, *Scotland's Relations*, 119.

[21] Seton-Watson, *Nations and States*, 7, counts them among the nine 'old continuous nations'.

[22] A. M. Duncan, *The Nation of the Scots and the Declaration of Arbroath (1320)* (London, 1970), 36. See also C. Harvie, *Scotland and Nationalism: Scottish Society and Politics 1707–1977* (London, 1977), 22–3.

[23] This proto-nationalism can be considered 'the minimal sense of cohesion and distinctiveness among a people'. Deutsch, *Nationalism and Social Communication*, 173.

Highlands, the area that represented the single greatest
challenge to national unity in every way, was the church
unable to establish the type of control that it exercised
elsewhere in the kingdom.

In 1600, therefore, both England and Scotland were, by
most definitions, national states, and the sense of communal
identity that characterizes nations appeared to be growing in
each. The prospect of 'perfect union' that emerged after the
Union of the Crowns, however, put the future of both national
states very much in question. For not only did King James
and many of those who espoused the cause of union speak in
terms of creating a new British state, but they also envisaged
the building of a new British nation. In fact, the impulse
towards nation-building was far stronger, especially during
the early seventeenth century, than the desire to forge a new
British state. Unionists like Sir Thomas Craig, Sir Henry
Savile, and after 1607 even James himself, all of whom expressed serious reservations about completely merging the
administrative structures of the two states, placed the highest
priority upon the establishment of a new nation. Many
unionists, of course, hoped that the new nation would serve as
the foundation for a new state. For them parliamentary and
administrative union remained a distant but real goal. To be
sure, if the union of the English and Scottish peoples was to
be genuinely national, it would by definition require some
political expression of that unity. But the immediate goal was
to facilitate the development of a British national consciousness, to create a new nation, just as other statesmen would
later create united Swiss, Spanish, German, and Italian
nations.

This was by all standards a bold enterprise. The boundaries
of nations, like the boundaries of states, could certainly be
changed. New nations could be made out of either heterogeneous ethnic groups, such as the Swiss, or relatively
homogeneous ones, such as the Germans. They could be made
by combining one national group, like the Castilians, with
non-national ethnic groups, such as the Valencians and
Aragonese. They could be made, probably most easily, by
separating one ethnic group from the state into which it had
been absorbed, as in the Netherlands. But to merge two well-

established nations such as England and Scotland, each of which possessed its own state sovereignty, required prodigious efforts at nation-building. As we shall see, those efforts failed in the seventeenth century, and their failure meant that the British state created in 1707 embraced not one nation but two. Once the new British state was established, however, the possibility arose that its institutions would help to build a British nation. That possibility never materialized, however, so that even today Britain remains a multi-national state.[24]

II

Soon after the Union of the Crowns James VI and I began to speak the language of national union. In his proclamation of May 1604 he commanded the subjects of both countries not only to consider the two realms united but to consider themselves as 'one people, brethren, and members of one body'.[25] One year later, when it had become clear that few of his subjects in either kingdom had heeded his command, he used even stronger language. 'He that doth not love a Scotsman as his brother', said James, 'or the Scotchman that loves not an Englishman as his brother is a traitor to God and the King.'[26] The threat may have been as idle as the command had been wistful, but the statement reflected an awareness, shared by most of the early unionist pamphleteers, that if a durable union was to be established, it would have to involve a 'union of love', a genuine recognition by the subjects of both kingdoms that they belonged to one nation.[27] Without such a national union a political, legal, or constitutional union could not survive. The apparently successful precedent for such a union of love was the union of the Welsh and English people after the constitutional union of 1536 and 1543.[28]

The enormous amount of planning and sheer energy that

---

[24] Rose, 'The United Kingdom as a Multi-National State'.

[25] *Stuart Royal Proclamations*, i. 19.

[26] SP 14/8/93.

[27] See for example Hayward, *Treatise of Union*, 8; Thornborough, *Discourse*, 17.

[28] Even before the Union of the Crowns the Earl of Northumberland suggested that James follow the Welsh example in this regard. See *Correspondence of King James VI of Scotland with Sir Robert Cecil and Others in England during the Reign of Queen Elizabeth*, ed. J. Bruce (Camden Society, 1861), 55–6.

James put into the building of a British nation has not been
fully appreciated. No British king or statesman of the seventeenth or eighteenth century was as thorough, imaginative,
or genuinely dedicated to the creation of a united national
community or British people as James. Had he succeeded—
and here it must be stressed that his failure was only in
small part his own responsibility—he would probably and
deservedly be counted among the great nation-builders of the
early modern period. There is no question that his task was
formidable. The Scots and the English inhabited the same
island, spoke similar languages, shared a common Protestant
faith, and were governed by many similar laws and institutions,[29] but the cultural and political barriers that separated
them in 1603 were far from insignificant. These barriers, the
memory of the bitter conflict that had taken place between
the two states in the past, and continued tension in the
Borderlands led Englishmen and Scots in 1603 to regard the
members of the other nation as foreigners. Changing these
habits of thought would tax the abilities of even the most
talented monarch.

James undertook this task with vigour, optimism, and
enthusiasm. In a conscious and deliberate fashion he tried
to effect an integration of the two peoples. This he sought
to achieve mainly within the context of a merely regal union.
It is true that James expressed a hope, at least in 1604, for
the eventual formation of a British state, an institution that
would be characterized mainly by a common parliament and
a common body of public law. James realized, however, that
the establishment of such a state would not be possible within
the immediate future, and therefore he decided to pursue the
goal of national union while maintaining two separate states
and parliaments.[30] The parliaments of the two countries
would of course be called upon to facilitate the process of
national integration, but the formal structure of the national
union would be that of a personal, dual monarchy. And once
the English parliament refused to ratify most of the provisions

---

[29] Donaldson, 'Foundations of Anglo-Scottish Union', 282–314.

[30] In 1607 he still spoke of a 'perfect union of laws and persons' but by then he
had abandoned hope of achieving the former and concentrated on the latter. See
Constitutional Documents, ed. Tanner, 35.

of the Instrument of Union of 1604, the motor force of national unification would be the resources of personal kingship—the allegiance owed to the king by all his subjects and the political powers he held by virtue of the prerogative. On the one hand the decision James made to pursue his goal in this way reflected his political realism, for the formation of a British state in 1607 would have been premature. On the other hand the decision continually hampered his efforts, since the monarchy did not have adequate resources to allow James to achieve his goals by himself.

Two of James's policies that were intended to bring about national unity—the achievement of religious conformity and the establishment of freedom of trade—have been discussed above. Both policies were intended, at least in part, to develop among Scots and Englishmen the belief that they belonged to the same community of people. Of the two, the policy of religious conformity had the greater potential, not only because the practice of religion was a more powerful source of cultural identification than the conduct of trade, but because the establishment of religious unity was more likely to have an impact on a larger number of people than the freedom to transport goods from one country to another. If James had been able to secure voluntary and widespread acceptance of basically English forms of worship and church government in Scotland, he might have established a sense of religious community similar to that which English puritans had with Scottish Covenanters in the late 1630s and 1640s, and thereby might have laid the basis of a powerful British nationalism. As it was, James did not achieve as much success as he wished in this regard, while his son failed ignominiously. Their successors in the late seventeenth and eighteenth centuries were left, therefore, with only a minimal sense of Protestant unity to serve as the basis of a potential British national identity. Pamphleteers in the early eighteenth century hoped that this sentiment could be successfully exploited for national ends,[31] but this proved to be possible only in times of international crisis.

The free trade that James tried so hard to establish also

[31] See for example Pyle, *National Union a National Blessing*, 17–18; [Defoe], *A Discourse upon an Union*, 13.

served, in a more indirect and subtle way, the purposes of national unity. Although it is true, as Jacob Viner has written, that a customs union cannot by itself make a fatherland,[32] it can help to create a sense of community among merchants. It can also help to establish a common economic life, which has often been recognized as one of the prerequisites of national unity. James's plans for commercial union, many of which were substantially blocked by the failure of the Instrument of 1604, were intended to achieve these effects. They also promised, in a symbolic but none the less meaningful way, to 'remove the mark of the stranger' by eliminating the 'alien customs' on Scottish goods imported into England and by treating Scottish ships as if they were English bottoms.[33] Whatever success James realized in this regard was reversed in the late seventeenth century, when the English parliament specifically excluded Scotland from participating in English economic activity. Only in 1707 was the economic foundation of a British national consciousness laid, and at that time it proved to be insufficient to the task.

In addition to pursuing the goals of religious and economic union, James realized that he had to take more direct action to remove the mark of the stranger that both Scots and Englishmen bore in the others' country. The first and most important step he took to achieve this end was to obtain mutual naturalization. Originally he hoped to do this much more expeditiously than in fact he was able to do. The Instrument of Union of 1604, to which James gave his full support, provided not only for a recognition of the mutual naturalization of the *post-nati* but for the passage of acts in both parliaments naturalizing the *ante-nati* as well. Although the latter group would have been excluded from offices of the crown, judicial posts, and membership in parliament, acceptance of the Instrument would have brought about an immediate recognition of the mutual naturalization of all Scots and Englishmen. As it was, James had to settle for the naturalization of only the *post-nati*, and this meant that the anticipated increase in British national awareness as the result of naturalization would be delayed for at least a generation.

[32] J. Viner, *The customs Union Issue* (London, 1950), 103.
[33] Lythe, 'Union of the Crowns', 22; Defoe, *History of the Union*, 722.

The means by which naturalization was achieved in England, the decision of the judges in *Calvin's Case* (1608), became the foundation of the English law of naturalization for the following two centuries.[34] By claiming that allegiance was personal rather than legal, to the king rather than to the law, it insured that individuals born in any of the king's dominions, even those who would be born overseas, would possess British citizenship. The theory of government upon which this is based is quintessentially feudal and 'medieval'. By refusing to separate the king's two bodies—the body politic and the body natural—and by insisting that a subject's allegiance was that of a man to his lord or a child to his parent, the judges perpetuated the old idea that subjectship was a natural, personal, and perpetual relationship. The view held by various opponents of naturalization that a man's allegiance was to the laws rather than the king was in many ways more modern, for it posited a more contractual relationship between the individual and the state. The judges, however, did not subscribe to this alternative view, and their decision helped to prevent the emergence of a modern legal definition of citizenship in Britain for the next two centuries. Indeed, it was only in the American colonies at the time of the Revolution that ideas of contract and consent successfully challenged the old notion of subjectship.[35] A similar development might have occurred in England at the time of the Republic, especially since that period witnessed the emergence of a concept of the 'citizen',[36] but the Republic did not last long enough for such ideas of citizenship to affect the legal framework within which allegiance, subjectship, and alien status were defined.

Calvin's Case had both a positive and a negative effect on the cause of Anglo-Scottish union. It laid the foundation for the possible union of England and Scotland into one nation but at the same time it postponed indefinitely the union of their laws and their incorporation into one state. If the judges

[34] W. Holdsworth, *A History of English Law*, ix (London 1926), 72–86.
[35] Kettner, *American Citizenship*, 131–209.
[36] Hanson, *From Kingdom to Commonwealth*, 3–40; M. Walzer, *The Revolution of the Saints* (Cambridge, 1965), 1–21; J. G. A. Pocock, *The Machiavellian Moment: Florentine Political Thought and the Atlantic Republican Tradition* (Princeton, NJ, 1975), 361–400.

had subscribed to the theory that allegiance was to the law
rather than the king, then the cause of naturalization would
have become inextricably linked with that of political and
legal union. In such a situation the desire of unionists for
naturalization would have increased the pressures in favour of
an incorporating union. By minimizing the ties of the subject
with the territorial state and by emphasizing a 'community of
allegiance' that transcended political boundaries and legal
jurisdictions,[37] the judges side-tracked the entire issue of
legal and political union while allowing James to continue his
efforts to bring the two peoples together in one community.

As it was, the mutual naturalization of Scots and
Englishmen born after March 1603 proved to be one of the
least expeditious methods of cultivating a sense of British
national identity. To suggest that it 'would in the course of a
few years produce the union automatically', as the Venetian
ambassador predicted in 1607, was simply wrong, as later
events proved.[38] Not only would the policy take some seventy
years to become fully effective, but if not reinforced by other
means it would remain an empty legal gesture. To comple-
ment the policy of naturalization, therefore, James sought to
integrate the adult, *ante-nati* populations of the two countries
in a variety of ways.

For various political and social reasons James's policies
of integration were highly restrictive. He had no intention of
encouraging an exodus of poor, lower-class Scots to England,
and he also wished to control the emigration of the Scottish
aristocracy.[39] Nevertheless James did begin the process of
creating a genuinely British court. Political considerations
prevented him from appointing all but a few Scots, such as the
Earl of Dunbar, William Young, and Sir James Creighton, to
offices of state, but he did grant posts in the household to a
total of 149 of his fellow countrymen.[40] James also included
Scots within his small circle of personal advisers and inaugurated the policy, further developed by his son, of appointing a

---

[37] Kettner, *American Citizenship*, 3–4, 28.

[38] *CSP Ven, 1607–10*, 485.

[39] *RPCS* viii. 655; ix. 173–4; Insch. *Scottish Colonial Schemes*, 57–8.

[40] Seddon, 'Patronage and Officers', 293–305.

limited number of Scots to the English privy council.[41] Within the church there were parallel efforts at assimilation. The ordination of a small number of Scots in England and their subsequent employment in English benefices probably did not come about as the result of royal initiatives, but the appointment of Scottish clerics to two deaneries and numerous prebendaries certainly received royal approval.

In addition to making these secular and clerical appointments, James tried to unite the Scottish and English people in four other ways. First, he supported efforts to enable Scots to attend the two English universities. James almost certainly realized, as modern scholars have observed, that the universities could be effective agents of nation-building, for they helped to break down the provincial mentality of the ruling class.[42] Since the universities, as well as the Inns of Court, had contributed to the growth of English national consciousness, it followed that they could have played a similar role in the creation of a British nation. The problem, however, was that the colleges at Oxford and Cambridge, which had gained control over university admissions only in the 1580s,[43] were restricted by statute from admitting any foreigners, and the heads of the colleges gave little evidence of a desire either to violate the statutes or to work for their repeal.[44]

Recognizing these obstacles, James used all the means at his disposal to realize his goals. Since the colleges could legitimately claim that they had no authority to annul their ancient prohibitory statutes, he directed the Earl of Salisbury, who was Chancellor of Cambridge, to confer with Archbishop Bancroft, Chancellor of Oxford, to see if they could use their powers of visitation to declare the statutes null and void. He also threatened to use his powers of patronage within the

---

[41] *CSP Ven. 1603–7*, 33, 57; Aylmer, *King's Servants*, 19, 24–5.

[42] For a cautionary note regarding this contribution of universities see V. Morgan, 'Cambridge and the "Country" 1540–1640', in *The University in Society*, ed. L. Stone (Princeton, NJ, 1974) 184–5. See also W. Prest, *The Inns of Court under Elizabeth I and the Early Stuarts, 1590–1640* (London, 1972), 39.

[43] See E. Russell, 'The Influx of Commoners into the University of Oxford before 1581: An Optical Illusion?', *English Historical Review*, 92 (1977), 734–5, on the establishment of this control at Oxford.

[44] For a similar policy at Eton see *Life and Letters of Sir Henry Wotton*, ed. L. P. Smith (Oxford, 1907), ii. 368.

universities against the wishes of the colleges if the collegiate authorities proved to be intractable on this question.[45]

The heads of the colleges, however, could not be moved. In responding to Salisbury's overtures they took refuge not only in the statutory prohibitions against the admission of anyone born outside England but also in their financial inability to support any additional members. By appealing to financial exigencies, they shifted responsibility for the exclusion of the Scots to the fellows, a majority of whom had to approve any new expenditures. The fellows, said the masters, 'will be adverse and backward to any such good purpose as this, because whatsoever is this way to be allowed must of necessity be defalked from them'.[46] As a result of this collegiate intransigence and the clear preference of Scots to attend their own universities or those on the Continent, very few Scotsmen matriculated at Oxford and Cambridge during the seventeenth century.[47] A number of Scots, including the minister Walter Balcanqual, who attended the Synod of Dort, were incorporated at the English universities after receiving degrees in Scotland or elsewhere. Others, such as John Gordon, the cleric who wrote in favour of the union, and John Durie, the ecumenical Scots preacher, were given honorary creations.[48] These admissions, however, were strictly honorific and were just as commonly granted to foreigners, so they could hardly be considered the tools of nation-building.

A second policy of assimilation, the fostering of marriages between the Scottish and the English, did not achieve much more success than the attempt to open English universities to the Scots. James, who frequently used marital analogies to describe the union,[49] also realized that a 'union of love' between the subjects of his two kingdoms would not be complete without the actual mixture of their blood. Prior to

[45] SP 14/50/43.

[46] *The Egerton Papers*, ed. J. P. Collier (Camden Society, 1840), 444–5.

[47] John and J. A. Venn, *The Book of Matriculations and Degrees, 1544–1639* (Cambridge, 1913), 589. The number of Scots admitted before 1603 may have actually been higher. See Donaldson, 'Foundations of Anglo-Scottish Union', 296–9.

[48] Henderson, *Religious Life*, 82; A. Wood, *Fasti Oxoniensis* (London, 1813–20), ii. 311, 383, 420–1.

[49] See M. J. Enright, 'King James and his Island: An Archaic Kingship Belief?', *SHR* 55 (1976), 28–40. For other usages of marital language in connection with the union see Peck, *Northampton*, 190.

1603 Anglo-Scottish marriages had been rare, except in the Borderlands, where they took place frequently and illegally.[50] James hoped that the Union of the Crowns would lead to an increase in such marriages, and he went out of his way to encourage matches among the gentry of the two countries. Thus in March 1606 he wrote to Sir Hugh Bethell to express his support for the marriage of his daughter to the son of Sir William Auchterlony, a Scot, and in 1607 he arranged a match between James Hay, the future courtier and diplomat, and the daughter of Lord Dennye.[51] All in all, however, James could do little to achieve his goal, especially since he deliberately followed a policy of restricting the flow of Scots southward. Contemporaries complained of the number of Scots in England, but there were probably far fewer than believed, even among the upper classes.[52] With such limited interaction between Scots and English, the opportunities for arranging marriages were greatly restricted. The author of *Rapta Tatio* confidently predicted in 1604 that 'Many will be the marriages in time, to make our nations fully one',[53] but there were very few Anglo-Scottish marriages outside the Borders during the entire seventeenth century. When Thomas Campion, on the occasion of Lord Hay's marriage, praised James for preparing 'the high and everliving union 'tween Scots and English' and claimed that 'he that marries kingdoms, marries men', he was exaggerating the powers of nation-building that the king possessed.[54] Perhaps it would have been more appropriate if he had cited the old Italian proverb, 'Kings may wed, but kingdoms never'.[55]

Closely related to James's marital policy was his determination to fuse the aristocracies of the two countries by giving English titles of nobility to *ante-nati* Scots. Lord Hay, who was created Baron Hay of Sawley in 1615 and Earl of Carlisle in

[50] T. I. Rae, *The Administration of the Scottish Frontier, 1513–1603* (Edinburgh, 1966), 11; Donaldson, 'Foundations of Anglo-Scottish Union', 309.

[51] SP 14/19/1; T. Campion, *The Discription of a Maske . . . in Honour of the Lord Hayes and his Bride*, in *Campion's Works*, ed. P. Vivian (Oxford, 1909), 57–76.

[52] Fuller claimed that there were not fewer than 3,000 Scots in England in 1606. Harl. MS 6842, f. 1. In 1567 it was reported that there were only 58 Scotsmen in London. C. Rogers, 'Memoir and Poems of Sir Robert Aytoun', *TRHS* 1 (1872), 110.

[53] *Rapta Tatio*, sig. F2ᵛ.   [54] Campion, *Discription of a Maske*, 59.

[55] *The History of the Union of the Four Famous Kingdoms of England, Wales, Scotland and Ireland* (London, 1660), 16.

1622, was only one of the beneficiaries of this policy; Robert Kerr, Viscount Rochester and Earl of Somerset, was another. Taken together with the arrangement of Anglo-Scottish marriages, the granting of titles was expected to produce what the Earl of Salisbury referred to as a 'union of nobility'.[56] The government adduced ample legal precedents for its policy,[57] but the titles, instead of contributing to the cause of Anglo-Scottish union, actually worked against it. In 1621 twenty-six English peers, angered by the selection of three Scottish peers to lead the procession of earls at the opening of the English parliament, drafted a protest against the Scottish invasion of the English peerage.[58]

Concerning the fourth part of James's assimilation policy, the union of the military forces of the two countries, little is known. It does appear, however, that James viewed the stationing of both English and Scottish regiments in the Netherlands as a means of cultivating a sense of common nationality.[59] If in fact that was his goal, the policy made sense. There is no more effective means of fostering a sense of national cohesion and identity than through the army. Christopher Hill has argued that the process of English national unification, especially the effective incorporation of the North and West into the mainstream of national life, was brought to completion by the New Model Army.[60] It is possible that joint military operations between England and Scotland after 1603 and even during the English Civil War may have given soldiers a certain sense of unity.[61] Certainly by the early eighteenth century those Scots who served with the English overseas were very much in favour of union.[62] But

[56] SP 14/26/80.

[57] SP14/26/69.

[58] V. Snow, *Essex the Rebel: the Life of Robert Devereux, the Third Earl of Essex, 1591–1646* (Lincoln, Nebr., 1970), 105–6. The question of ranking Scottish peers had been a stumbling block to the negotiation of a union ever since 1604. See Bacon, 'Certain Articles', *L&L* iii. 230; *CSP Ven. 1603–7*, 148; Notestein, *House of Commons*, 212.

[59] *Political Works of James I*, 304. An additional motive, however, may have been to remove the Scottish 'rabble' from the kingdom. *CSP Ven. 1603–7*, 107–8.

[60] C. Hill, *The World Turned Upside Down (London*, 1972), 65.

[61] In 1603 Scottish troops came to the assistance of the English in a quarrel with the French in the camp of Count Maurice. *CSP Ven. 1603–7*, 87.

[62] S. H. F. Johnston, 'The Scots Army in the Reign of Anne', *TRHS*, 5th ser. 3 (1953), 20.

military co-operation could not foster a genuine sense of national cohesion unless the two military establishments were united, and that did not happen until 1708.[63] Even then, however, the maintenance of separate English and Scottish regiments inhibited the development of British national consciousness within the ranks.

In order to symbolize and to some extent even foster the process of Anglo-Scottish assimilation, James promoted the use of the term Great Britain to describe his united kingdom. His main objective in attempting to change the names of his two kingdoms by statute and then, when that failed, in adopting the new style by proclamation, was to celebrate his dynastic achievement. The routine use of the term Great Britain both at home and abroad would have served as a constant reminder of the permanent enlargement of the king's dynastic patrimony, just as the striking of a new British coinage and the creation of a new Union Jack for the king's ships called attention to the fact of regal union.[64] James hoped, however, that the new name would also contribute to the growth of a new national identity. His wish was that people would begin referring to themselves as Britons or at least as North Britons and South Britons. The problem was that the use of a new name to describe one's nationality usually reflects, rather than inspires, a new national identity. James had no reason to expect his subjects to consider themselves British rather than English or Scottish until they had already developed a sense of being primarily British for other reasons. And since the other aspects of James's plan had not met with overwhelming success, there was little hope that the terms 'Great Britain' and 'Briton' would acquire customary usage. In fact, the adoption of the name of Great Britain had been strongly opposed by the Scots and the English in 1604, while almost twenty years after the king had assumed the new style one pamphlet reported that 'they make a mock of your word "Great Britain"'.[65] The only evidence we have of its use, except

[63] J. M. Fortescue, *A History of the British Army* (London, 1902–30), i. 582; H. C. B. Rogers, *The British Army in the Eighteenth Century* (London, 1977), 18.

[64] *Stuart Royal Proclamations*, i. 7, 99–103, 135–6; NLS, MS 2517, ff. 67ᵛ–68.

[65] *CSP Ven. 1603–7*, 94, 106; SP 14/7/58; Harl. MS 292, ff. 133, 136; Bodl., Tanner MS 75, f. 44; 'Tom Tell-Troath' in *Complaint and Reform in England, 1436–1714*, ed. W. H. Dunham, Jr., and S. Pargellis (New York, 1938), 482. In 1707 Defoe was

in diplomatic communications and in official Scottish documents, comes from the letters of a few self-conscious unionists and the works of a number of poets, scholars, and dramatists.[66] As an experiment in nation-building the change in the royal style proved to be no more successful than the hollow and purely symbolic use of the Union Jack, which aroused more national antagonism than feelings of national unity.[67]

The policies that James followed with respect to the Borders, though in large measure intended to achieve the age-old goal of establishing law and order throughout the king's dominions, also served the purposes of national unification. The comparatively lawless state of the Borders stood as a reminder not only that the King's Peace was not being maintained in all his shires, but that English and Scots were still at war with each other. It is true of course that the thieving and murdering that took place on the Borders did not always involve English–Scottish hostilities. The Border reivers, as they were called, just as readily attacked their own countrymen as individuals across the Border.[68] Nor is it any more true that Scotsmen and Englishmen on the Borders were always hostile to each other, for many of the Borderland families had either kinship ties or formal alliances with families on the other side of the Border. One family, the Grahams, clearly benefited from the dual allegiance that such ties made possible.[69] What mattered most in the Borderlands was loyalty to the family, not to the nation or the state.[70] Never-

still talking about the assumption of the name of Britain as one of the effects of the union that was being negotiated. Defoe, *The True-Born Britain* (London, 1707), 4.

[66] Lodge, *Illustrations*, iii. 240; J. Speed, *Theatre of the Empire of Great Britaine* (London, 1611); M. Drayton, *Poly-Olbion*, in *The Works of Michael Drayton*, ed. J. W. Hebel (Oxford, 1933); L. Lloyd, *The Jubile of Britane* (London, 1607), sig. A2ᵛ. On the failure of the terminology to penetrate the vernacular see Wedgwood, 'Anglo-Scottish Relations', 33. See generally S. T. Bindoff, 'The Stuarts and their Style', *English Historical Review*, 60 (1945), 212–16; Lee, *Road to Revolution*, 105.

[67] *RPCS* vii. 498–9. The debate on the change of the name had a similarly negative effect. HMC, *Salisbury MSS*, xvi. 98.

[68] G. M. Fraser, *The Steel Bonnets* (New York, 1972), 8.

[69] See R. Newton, 'The Decay of the Borders: Tudor Northumberland in Transition', in *Rural Change and Urban Growth, 1500–1800*, ed. C. W. Chalklin and M. A. Havinden (London, 1974), 7.

[70] G. M. Trevelyan, 'The Middle Marches' in *Clio, A Muse and Other Essays Literary*

theless, Anglo-Scottish rivalry, which had originally been responsible for giving Border life its harsh and warlike character, still played a major role in the robbing and blackmail that so frequently occurred in the region. Indeed, one of the most distinctive features of Border justice in the sixteenth century were the 'days of truce', the meetings of the wardens from both sides of the Border to settle the numerous complaints made by Englishmen and Scots against the raiding parties of the other nation.[71] Although these meetings were intended to promote harmony between two nations that were formally at peace with each other, they often served the opposite purpose. Unable to allay the suspicion of one side that appropriate justice was not being administered by the other, the days of truce often proved to be tense and occasionally violent encounters.

The determination with which James undertook the pacification of the Borders after the Union of the Crowns can only be explained by his commitment to the cause of a durable national union. Open defiance of the law in any part of his two kingdoms would certainly have concerned him greatly, but the situation in the Borders represented more than mere disregard for the law. Raiding parties crossing the Border stood as reminders that the frontier between his two kingdoms had not disappeared and that his English and Scottish subjects still regarded the others as foreigners and enemies. Tension between English and Scots on the Borders remained so strong four months after the Union of the Crowns that the Venetian ambassador doubted whether the two nations could ever 'pull together'.[72] The Borders, therefore, called for special attention, and this they certainly received. On the one hand James, who now had the capacity as king of both nations to ensure co-operation between Scottish and English authorities and thus to prevent criminals from playing off one side against the other, established what one writer has referred to as a 'police state which was most barbarously administered'.[73] On the

---

and Pedestrian (London, 1913), 164. Trevelyan was wrong to see family allegiance as the only allegiance.

[71] Fraser, *Steel Bonnets*, 153–67.
[72] *CSP Ven. 1603–7*, 70–1.
[73] Fraser, *Steel Bonnets*, 362.

National Union

other hand, he attempted in a number of less Draconian ways to foster a genuine sense of national union. These included the designation of the Borderlands as the Middle Shires, the appointment of a joint Border commission under the leadership of the Earl of Dunbar, and the establishment, after a long parliamentary struggle, of the practice of remanding prisoners across the Border for trial in the country where they had committed their crimes.[74]

In evaluating the success of James's Border policy, we must distinguish between its strictly judicial and its general unionist objectives. As far as the judicial policy of pacification was concerned, the king must be credited with a fair measure of success.[75] The Borders were not completely tamed until the eighteenth century, but a new and lasting peace had dawned before James died. Sir Edward Philipps may have spoken somewhat prematurely when he claimed that the Borders were as free of crime as any other part of England in 1609, but by that time the commissioners had already made great progress in destroying the 'system of the frontier' and in reducing the level of criminal activity. The days of the Border 'riding families', of interminable, sustained family feuding, and of collective violence had passed.[76]

As far as the union was concerned, however, James achieved much less success. The designation of the Borders as the Middle Shires gained even less currency than the term Great Britain with which it was closely linked. Residents of the Border shires, who were as genuinely 'British' by blood and geography as any other group of Englishmen or Scots, remained far too preoccupied with their own familial and regional identity and too conscious of the significance of the Borders in their life to become pioneers in the process of national redefinition. The Border commissions themselves, however effective they may have been in reducing the level of violence in the Borderlands, did not encourage the type of

---

[74] 7 & 8 Jac. 1, c. 1, in Tanner, *Constitutional Documents*, 43–5.

[75] Gay, 'Border Commissions', 194. Gay emphasizes more than other writers the persistence of violence in the Borderlands.

[76] Fraser, *Steel Bonnets*, 374; P. Williams, 'The Northern Borderland Under the Early Stuarts', in *Historical Essays 1600–1750 Presented to David Ogg*, ed. H. E. Bell and R. L. Ollard (London, 1963), 10, 14–15; R. Millward, 'The Cumbrian Town between 1600 and 1800', in *Rural Change and Urban Growth*, ed. Chalklin and Havinden, 203–7.

harmony that would advance the cause of union, nor did they lay the foundation of durable British institutions, as Bacon had hoped they would in 1604.[77] The Border as an armed frontier may have disappeared in the early seventeenth century,[78] but the Border as the line separating two states and two nations remained.

In explaining the failure of James to achieve the union of love that he so ardently desired, it is easy to place the blame on the king himself. James has become the butt of so much historical criticism that censure of this sort suggests itself almost instinctively. The criticism is in large measure misplaced. There is of course no question that James made his fair share of blunders in trying to achieve union. In promoting the union of love he often acted precipitately and in some ways actually exacerbated rather than reduced Anglo-Scottish tensions. But the reasons for James's inability to build a British nation lie much more in the 'mutual repugnance' Englishmen and Scots had of each other than in the tensions that James created.[79] Even if James had proceeded more cautiously in his pursuit of perfect union, he would still have encountered enormous difficulties in getting the subjects of his two kingdoms to regard each other as countrymen and brethren.

One of the main sources of animosity between Englishmen and Scots was the memory of previous armed conflict between them. Less than sixty years before the Union of the Crowns England and Scotland had been at war with each other, and as modern experience amply testifies, the hatred of national enemies can often persist long after the conclusion of peace and even after the establishment of alliances between them. In the case of England and Scotland, the continued operation of the hostile laws during the Elizabethan period (which were not repealed until 1607), the persistence of conflict on the Border, the continuation of a special relationship between Scotland and France after 1558, and tensions that arose over the execution of Mary Queen of Scots[80] all helped to

[77] Williams, 'Northern Borderland', 9; *L&L* iii. 221.
[78] This was symbolized by the dismantling of the English garrison at Berwick. See Watts, *From Border to Middle Shire*, 311.
[79] P. H. Brown, *History of Scotland* (Edinburgh, 1902), ii. 248.
[80] *The Memoirs of Robert Carey*, ed. F. H. Mares (Oxford, 1972), 7–8.

perpetuate the old antagonisms. The French ambassador in England claimed that 'rooted and ancient hostility' of the English to the Scottish was so strong in 1603 that it threatened to retard the king's arrival in England, while Contarini, the Venetian ambassador, reported that Scottish opposition to the departure of the king arose 'out of hatred for the English'.[81] According to an English writer in 1604, 'the continued unkindness (a smoother phrase I cannot use) between us and Scotland for hundreds of years past' still lived on in the hearts of 'the worser sort'.[82]

A second major source of tension between Englishmen and Scots was the attitude of cultural and economic superiority that the English developed with respect to their northern neighbours. These sentiments, which predate the Union of the Crowns, find much more frequent expression after that date, since it was only then that the possibility of large-scale social intercourse arose.[83] Fearful of an invasion of lean Scottish kine into their green pastures, Englishmen frequently uttered disparaging statements about the Scots, depicting them as poor, unclean, uncivilized boors, who had a penchant for begging, lying, and violent crime.[84] As Francis Osborne wrote in his 'Traditional Memoirs',

> They beg our lands, our goods, our lives
> They switch our nobles, and lie with our wives;
> They pinch our gentry, and send for our benchers,
> They stab our sergeants and pistol our fencers.[85]

The image of Scots as criminally dangerous received reinforcement from their record of apparent political instability. As Sir Christopher Piggott remarked in the English

---

[81] *CSP Ven. 1603–7*, 15, 16.

[82] SP 14/10A/17, f. 5.

[83] J. Wormald, 'Gunpowder, Treason, and Scots', *Journal of British Studies*, 24 (1985), 159–60; H. J. Hanham, *Scottish Nationalism* (London, 1969), 73–4. For the statement of Thomas Fuller in 1606 see Harl. MS 6842, f. 1.

[84] *Secret History of the Court of James the First* (Edinburgh, 1811), i. 217–40; Sir A. Weldon, 'A Description of Scotland', in Nichols, *Progresses*, iii. 338–43; G. Goodman, *The Court of King James the First*, ed. J. S. Brewer (London, 1839), 320–1; C. H. Firth, 'The Ballad History of the Reign of James I' *TRHS*, 3rd ser. 5 (1911), 23–4; *Bowyer Diary*, 203–4; *RPCS* vii. 402 n.

[85] *Secret History of the Court of James the First*, i. 217. For an echo of these charges in the 1640s see *Diary of Burton*, ii. 384 n.

parliament of 1607, the Scots were not only thieves and murderers but also regicides.[86] The composite impression that all these national slurs made was that the Scots were a poor and uncivilized people with whom it would be both unprofitable and dangerous to associate.[87] An Englishman predicted in 1604 that the poor Scots would 'flock hither in such multitudes as that death and dearth is very probably to ensue'.[88] Exploiting this anti-Scottish sentiment, George Chapman, Ben Jonson, and John Marston inserted in their comedy *Eastward Ho* (1604) a proposal for shipping 100,000 Scots to the New World, where 'we should find ten times more comfort of them . . . than we do here'.[89]

At the same time that Englishmen were expressing and developing their contempt for poverty-stricken, base Scots, they were resenting the favouritism that the king was showing to the select group of Scots who found places in his court. The problem was twofold. On the one hand, the king was granting money, annuities, and according to one report, even gold chains to his Scottish cronies; on the other hand he was allowing them to run his government.[90] There was a solid foundation to the first charge. During the first few years of his English reign, James bestowed upon Scots £10,614 in pensions, £88,280 in 'ready money', £133,100 in old debts, and £11,093 in annuities.[91] The second charge was somewhat exaggerated, for as mentioned earlier, James had made very few appointments of Scots to offices of state. Nevertheless, James had made himself unusually accessible to his Scottish associates and thus gave the impression that these men were exercising undue influence in the formulation of royal policy, so much so that contemporaries referred to the 'Scottish government' of England.[92]

Resentment against the Scots for their preferential treatment mounted steadily during the early years of James's

[86] Cobbett, *Parliamentary History*, i. 1097.
[87] Even arson in the king's stables was attributed to them in 1609. HMC, *Salisbury MSS*, xxi. 96.
[88] SP 14/10A/17, f. 4.
[89] *Eastward Ho*, ed. R. W. Van Fossen (Manchester, 1979), 139.
[90] Bodl., Carte MS 82, f. 157ᵛ; 85, f. 23; 86, ff. 473–474.
[91] Add. MS 12497, ff. 153–160.
[92] *CSP Ven. 1603–7*, 33–7; G. Holles, *Memorials of the Holles Family, 1493–1656*, ed. A. C. Wood (Camden Society, 1937), 94–5; Bodl., Carte MS 82, ff. 160ᵛ–161.

reign, and it became a major grievance in both the parliamentary session of 1610 and the parliament of 1614.[93] Towards the end of the former assembly MPs identified the Scots as the source of the crown's financial problems and, according to the French ambassador, 'wished *mal de mort*' to them.[94] In 1614 John Hoskyns, who had spoken against the Scots in 1610, stated that a wise prince would send the Scots home, just as Canute had done with the Danes who had followed him to England. Hoskyns also made a threatening reference to the Sicilian Vespers, the rebellion against Angevin domination of Naples and Sicily in 1282.[95]

To all these expressions of anti-Scottish sentiment the Scots responded in kind, expressing their hostility and 'insolency' towards the English. Both in speech and writing, it was reported, 'they slander, malign, and revile the people, estate, and country of England'.[96] Just as English speeches in the House of Commons 'roused disgust and anger in the breasts of the Scottish', so the hostile remarks of the Scots threatened to 'stir up' in the English 'irreconcilable evil will'.[97] Instead of bringing about a union of love, the prospect of union had ironically revived 'the ancient enmity between the two countries'.[98] In more than one instance this enmity took the form of duels between the Scots and Englishmen at court.[99]

James expressed deep concern over the frequent expression of anti-Scottish sentiment in England and its converse in Scotland, but there was little he could do to suppress it. On the advice of the Earl of Dunbar he had the Scottish parliament pass an act prohibiting all anti-English writing that was harmful to the union, while in England he prosecuted the authors of *Eastward Ho* and brought about the deletion of

[93] 'Advertisements of a Loyal Subject to his Gracious Soveraign', in *Somers Tracts*, ii (1810), 144–8; Bodl., Carte MS 105, ff. 93–95; *Proceedings in Parliament, 1610*, ii. 344–5; HMC, *Salisbury MSS*, xxi. 263.

[94] *Proceedings in Parliament, 1610*, ii. 345 n.

[95] T. L. Moir, *The Addled Parliament of 1614* (Oxford, 1958), 138.

[96] *APS* iv. 436. See also *RPCS* vii. 356; *Stuart Royal Proclamations*, i. 39.

[97] *CSP Ven. 1603–7*, 153; *APS* iv. 436.

[98] Ibid. 485.

[99] Sir B. Whitelocke, *Memorials of the English Affairs . . . to the End of the Reign of James I* (London, 1709), 287; *Secret History of the Court of James the First*, i. 227. See Craig, *De Unione*, 468, for an effort to prevent duels 'undertaken to maintain the dignity of either kingdom'.

the offending lines.[100] In dealing with the English parliament he combined defences of his policies with stern warnings against the disparagement of the Scots, claiming that the Scottish nation 'cannot be hated by any that loves me'.[101] His reaction to the speech by Piggott was so strong that parliament 'was obliged' to punish him.[102] In these ways James was able to contain the public expression of the antagonism that prevailed between the two countries, but he could not remove the sentiment itself, even if he had agreed to send the Scots home. In the final analysis James was faced with the fact that the two nations he wished to fuse into one were not ready to be merged.

III

After the early years of James's English reign very few conscious efforts were made to achieve the type of national union that James had envisaged. The monarchs who succeeded him dealt frequently with the difficult problem of ruling the two nations jointly, and they occasionally supported projects for a further union, but they gave little evidence of an interest in uniting the 'hearts and minds' of the English and Scottish people. Their reluctance to engage in a process of nation-building may have stemmed from a recognition of the futility of any such effort, but it is more likely that they considered it unnecessary. Not having shared James VI's experience of ruling Scotland by itself, they never acquired the paternal attitude towards their Scottish subjects that direct and exclusive rule had encouraged in their predecessor. For Charles I and his successors, Scotland was a distant state that needed to be ruled as effectively as possible, not a nation to be joined in a bond of love with the English people. Union, even an incorporating union, would be a worthwhile goal if either English interests or the security of Britain required it, but

---

[100] *APS* iv. 436; Lee, *Government by the Pen*, 83; *Eastward Ho!*, 4–8.

[101] HMC, *Salisbury MSS*, xxi. 265.

[102] *CSP Ven. 1603–7*, 478–9. The speech that sent the king into a 'violent fit of anger' is almost certainly that of Piggott, not 'Deigus'. See Bodl., Calendar of Carte MS, vi. 361. Piggott was released on a plea of ill health, mainly because he could not legally be imprisoned while parliament was in session. *Spain and the Jacobean Catholics*, ed. Loomie, 97.

national union had little to recommend it. Even the practical desire of James VI and I for national union as a means of strengthening the regal union held little appeal, mainly because the regal union did not appear to be threatened until the early eighteenth century. When that threat did arise, the remedy took the form of merely an incorporating union—the union of the two states—not the union of two peoples.

The only time that Englishmen and Scots spoke the language of national union after the death of James VI and I was during the 1640s, when commissioners from both countries conducted a series of negotiations in connection with the Bishops' War and the English Civil War. The papers submitted by the Scottish commissioners before the Truce of Ripon in 1640 and the Treaty of London in 1641 were the first of many proposals for a genuine union of the English and Scottish people. The main concern of the Scots in these negotiations was, of course, religious union, not the type of liturgical uniformity that Charles had tried to impose in the 1630s, but a religious covenant that would involve the two nations in a union of hearts and minds. They also sought a commercial union, a proposal which reflected their obvious self-interest but which also would have strengthened the union of the two peoples in a tangible way. Their overall goal was to produce a 'mutual communication betwixt the two Nations'.[103]

There is plenty of evidence that the Covenanters had in mind something more than mere religious union. As early as 1640, when the Scots published a statement regarding their objectives in entering England, they expressed the hope that the two kingdoms would live in greater love and unity than ever before.[104] In 1644, when issuing a similar statement, they claimed that their purpose was 'that we may not be looked with the prejudice of strangers, which we hope the firm union of this mutual Covenant will wear out'.[105] That same year Samuel Rutherford, pleading for help for the 'bleeding sister church', claimed that their solidarity, which was now

---

[103] Harl. MS 455, f. 23ᵛ.

[104] 'Information from the Scottish Nation', in *Notes of the Treaty Carried on at Ripon*, ed. J. Bruce (Camden Society, 1869), 70–1.

[105] *A Short Declaration of the Kingdom of Scotland for Information and Satisfaction to their Brethren of England* (Edinburgh, 1644), 5.

embodied in the Solemn League and Covenant, had more than a religious foundation. 'We sail in one ship together', wrote Rutherford, 'being in one island, under one king, and now by the mercy of God have sworn one covenant and must stand or fall together.'[106] There is little doubt that Rutherford was thinking of a united British people. In like manner a number of Covenanters who abandoned the cause and made peace with Charles in the Crown Petition of 1643 referred to themselves as 'we British subjects'.[107]

The English did not pursue the goal of union with the same degree of enthusiasm as the Scottish Covenanters, and when they signed the Solemn League and Covenant in 1643 they did so mainly to secure Scottish military assistance in the Civil War. In their dealing with the Scots during these years, however, the English did often use the language of national union, and they took steps to counteract the persistent anti-Scottish sentiment that had plagued the nation-building efforts of James I. The advocates of a Scottish alliance, such as Simonds D'Ewes and John Pym, discussed the relationship between the two nations in a manner reminiscent of James and his literary supporters. When D'Ewes proposed both negotiations with the Scots and the granting of assistance to them, he insisted that the Scots were not enemies of the English but a people united with them under one monarch. 'We and they were originally one nation', said D'Ewes, 'sprung and descended from one and the same people and speaking the same language, having only different names. And through God's goodness as we were branches of one root in the beginning, so now we were subjects under one monarch.'[108] Attacks on Charles I and his councillors stressed the same theme. One of the grievances that Pym listed in 1640 was that the 'popish' forces had attempted 'to make a difference between England and Scotland'.[109] A further sign of English efforts to foster national union with the Scots was the

[106] [S. Rutherford], *Lex Rex: The Law and the Prince* (London, 1644), 381–2.
[107] E. J. Cowan, 'The Union of the Crowns and the Crisis of the Constitution in 17th-Century Scotland', in *The Satellite State in the 17th and 18th Centuries*, ed. S. Dyrvik *et al.* (Bergen, 1979), 131.
[108] *The Journal of Sir Simonds D'Ewes from the Beginning of the Long Parliament to the Opening of the Trial of the Earl of Strafford*, ed. W. Notestein (New Haven, 1923), 320.
[109] Ibid. 9.

reprinting of John Thornborough's two strongly pro-union treatises of 1604.[110] Thornborough, an English bishop, would hardly have approved of the religious union envisaged by English puritans and Scottish Covenanters, but the language that he and the puritans of the early 1640s used to describe the union of the English and Scottish nations exhibited striking similarities.

The efforts of the English and Scottish commissioners to strengthen the bond between the two nations could not, however, overcome the deep distrust and overt hostility that each nation had of the other. The most virulent expression of this nationalist sentiment came from the English royalists, who were naturally eager to disparage the Scottish rebels. The poems of John Cleveland, who more than any other royalist writer helped to discredit the Scots in English eyes, exploited the traditional English contempt for their northern neighbours.[111] His couplet, 'Had Cain been Scot, God would have changed his doom / Not made him wander but compelled him home', recalled earlier references to the poverty and filth of Scotland, while his characterization of the Scots as rebels and assassins of monarchy harked back to one of the themes of anti-unionist speeches in the English parliament of 1607.[112] Other royalist ballads appearing in the 1640s labelled the Scots as pedlars, just as English MPs had described them in the early seventeenth century.[113]

If all the sources of this anti-Scottish invective had been royalist, the damage to the cause of national union might not have been so devastating. As it was, distrust and open contempt for the Scots also surfaced within the parliamentary camp, especially after 1644.[114] Criticized for military ineffectiveness, and resented for their intervention in English politics, the Scots faced increasing animosity from the very

---

[110] *The Great Happiness of England and Scotland by being Re-united into One Great Britain* (London, 1641).

[111] Nevo, *Dial of Virtue*, 27.

[112] Firth, 'Ballads Illustrating the Relations of England and Scotland', 116; Nevo, *Dial of Virtue*, 58.

[113] Firth, 'Ballads Illustrating the Relations', 117. See also J. Tatham, *The Scots Figgaries: or, a Knot of Knaves. A Comedy* (London, 1652), 1.

[114] V. Pearl, 'Oliver St. John and the "Middle Group" in the Long Parliament: August 1643–May 1644', *English Historical Review*, 75 (1966), 498.

parliament with which they were allied. The presence of Scottish troops in England gave rise to 'jealousies . . . and misunderstandings betwixt us and the English', while the Scottish control of the royal person in 1646 greatly damaged whatever standing the Scots still had in English eyes.[115] Underlying much of this parliamentary estrangement from Scotland was dissatisfaction with the Covenant itself. Accepted primarily as a condition of Scottish military assistance, the Covenant stood as a challenge to the Erastian predilections of the great majority of English MPs and the ecclesiastical preferences of all but a small group of English presbyterians. One 'English Covenanter', writing in 1648, assigned the label of 'Independency' to Scottish presbyterianism and attacked it in these words:

Truly Sirs, your Scotch Independency is as distasteful to us, as that in England or Amsterdam. If you say it is the COVENANTED Religion, according to the word of God, and the example of the best reformed churches, we wait with patience and hearty prayers and hands lifted up to the most high God for the speedy settling of that amongst us; but assure yourselves, except you will impose your Scottish sense upon our English words (intolerable slavery) we resolve to be, according to our Solemn League and Covenant, English Presbyterians and not Scottish Independents.[116]

Despite the strength of all this anti-Scottish sentiment, the Covenanters did not abandon their pursuit of the goal of religious and national union. In 1646 the Marquis of Argyll appealed to the English parliament to 'hold fast that union which we happily established betwixt us, and let nothing make us again two, who are so many ways one; all of one language, in one island, all under one king, one in religion, yea, one in Covenant, so that in effect we differ in nothing but in the name (as brethren do), which I wish were also removed that we might be altogether one, if the two kingdoms shall think fit'.[117] Even when the rise of the New Model Army and the English Independents, who despised the Scots, sealed the fate of both the Covenant and the Scottish-parliamentary

---

[115] SRO, PA 7/24, f. 215; L. Kaplan, *Politics and Religion during the English Revolution: The Scots and the Long Parliament* (New York, 1976), 120.

[116] *The Scottish Mist Dispel'd* (London, 1648), 2–3.

[117] *LJ* viii. 392–6.

alliance, the Covenanters did not abandon their unionist goals. Having decided to throw their lot in with the royalists and the English presbyterians against the Independents, the moderate Covenanters subscribed to the Engagement of December 1647, in which they articulated the same goals of religious, economic, and national union that they had espoused in the early 1640s.[118] Whether this unionism represented mere political posturing or a sincere restatement of their earlier dreams cannot be determined.[119] Even if the sentiment was genuine, however, it had little hope of establishing a binding national union between two people who had such little genuine affection for each other.

The Independents' victory over the Engagers in the second Civil War and the subsequent invasion of Scotland by the New Model Army spelled the end of all hope for national union in the seventeenth century. The military success of Cromwell in Scotland did, of course, lay the basis of an incorporating union in 1654, but that union did not in any way involve the fusion of the two nations. Whether Cromwell or his associates even considered the possibility of a national union is highly questionable. A number of historians have attributed to Cromwell a desire to eradicate all distinctions between the Scots and the English and thus to realize the dream of James VI and I.[120] To be sure, Cromwell did bring about a parliamentary and commercial union, and he hoped to unite English and Scots law. These policies were not designed, however, to promote a 'union of love', the acceptance of the people of both countries as brethren.[121] Intended mainly to provide security for England and to facilitate English rule of Scotland, they were the product of a state-building, not a nation-building mentality. Cromwell wished to incorporate 'all the people of Scotland . . . into one Commonwealth [i.e., state] with England', as the Ordinance of Union

[118] For the text of the Engagement see *Constitutional Documents of the Puritan Revolution*, ed. Gardiner, 347–52. See also the speech of Lauderdale in *Declaration of the Commissioners for the Kingdom of Scotland* (London, 1647), 6–8.

[119] Ferguson, *Scotland's Relations with England*, 131, 136, 290 n.

[120] Ibid., 138; i. Roots, *The Great Rebellion, 1642–1660* (London, 1966), 175.

[121] The only references to love between Cromwell and the Scots appeared in 1648. *A True Account of the Great Expressions of Love from the Noblemen, Ministers and Commons of the Kingdom of Scotland unto Lieutenant General Cromwell* (London, 1648); *The Clarke Papers*, ed. C. H. Firth (Camden Society, 1891–1901), ii. 152.

of 1654 proclaimed,[122] but he had no desire to establish a new British national identity. One could hardly have expected a man who in 1648 had referred to Scotland as a 'foreign nation' and who considered England to be the elect nation to pursue such a goal of national fusion.[123] Cromwell remained, until his final days, 'God's Englishman', being 'ever English of the English'.[124] Even the language of the Cromwellian Union betrayed the limited, essentially political nature of the conjunction he so carefully engineered. Eschewing all references to Britain or Great Britain, authorities referred simply to the 'Commonwealth of England, Scotland, and Ireland', a style that parliament itself adopted in 1656.[125]

Even if Cromwell had actually shared the genuinely British vision of James VI and I, the policies he pursued prevented the development of British national consciousness. Union or no union, Scotland remained throughout the Protectorate a militarily occupied country, a condition that hardly encouraged a Scottish view of the English people as their brethren. The garrisons that remained active until 1660 served as constant reminders to the Scots that they had in fact been conquered, even if that conquest had been followed by political incorporation. Whatever benefits English rule brought to the country—and in terms of administrative and judicial efficiency they were not inconsequential—were offset by the resentment engendered by the English military presence.[126] That resentment also prevented one of the social consequences of military occupation, the marriage of English soldiers to Scottish women, from serving the interests of national union. Instead of breaking down the social barriers that separated the two nations, these marriages threatened the normal relationship between occupier and occupied, between conqueror and conquered, and therefore were prohibited by the authorities.[127] In the Scottish kirk sessions, familiarity with

---

[122] *CSPD, 1654*, 90.

[123] *The Writings and Speeches of Oliver Cromwell*, ed. W. C. Abbott (Cambridge, Mass., 1937), i. 691–2.

[124] J. Morley, *Oliver Cromwell* (London, 1900), 255.

[125] *Diary of Burton*, iv. 99–100, 128.

[126] The only expressions of affection came from MPs at Westminster. See for example *Diary of Burton*, iv. 137–8.

[127] Firth, 'Ballads Illustrating the Relations', 122–3.

English soldiers constituted indisputable evidence of moral turpitude.[128]

In the final analysis the Cromwellian Union probably did more to exacerbate tensions between the English and Scottish people than to reduce them. Even the freedom of trade that the Union entailed, the one aspect of Cromwell's policy that might have fostered a national union, caused deep resentment among English merchants against the Scots. Once the Cromwellian Union came to an end, the prospects for national union became even bleaker. During the Restoration period religious and legal differences between the two countries became more profound, while their economic rivalry intensified. In such circumstances there was very little likelihood that the Scots and the English could ever view each other as members of the same nation. The union negotiations of 1668, 1670, 1702–3, and 1706 reflected this reality. In these negotiations the quest for national union that had inspired the efforts of both James I and the Covenanters was absent. The commissioners and statesmen who discussed the problem concerned themselves almost exclusively with the possibility of political and economic union. They debated the formation of a British state, not the building of a British nation.

IV

In light of the complete lack of concern for national union that characterized the union negotiations of the late seventeenth and early eighteenth centuries, the failure of the Treaty of Union of 1707 to create a British nation should not surprise us. The Treaty established a British state, nothing more and nothing less.[129] The use of the term Great Britain to describe this new state, so highly evocative of King James's union project, should not lead us into the common error of seeing the Treaty as the fulfilment of James's vision.[130] Not only was the

---

[128] SRO, CH 2/357/1, records of the Tranent kirk session, 15 Nov. 1656.

[129] The only effect that the Treaty had on the union of the two peoples is that both Englishmen and Scotsmen became citizens of Great Britain. This prevented England from considering Scots as aliens for commerical purposes. See Smith, *Studies Critical and Comparative*, 27; Kettner, *American Citizenship*, 48.

[130] Willson, 'Anglo-Scottish Unity', 54; D. Hay, 'England, Scotland and Europe: The Problem of the Frontier', *TRHS*, 5th ser. 25 (1975), 87.

British state of 1707 fundamentally different from the political formation envisaged by James, but it failed to inspire the type of British national consciousness that James had hoped to inculcate. Indeed, the very process of achieving the Union of 1707, which was marked by acrimonious political debate, a fierce pamphlet war, numerous popular protests against the union in Scotland, and even the fear of war between the two countries, did more to fuel national antagonisms than to allay them.[131] As W. S. McKechnie has written, 'In 1707 the sentiment of nationality was not a bond of union but rather a knife that cut the island sharply into hostile units.'[132] No less an advocate of union than Daniel Defoe admitted as much only a few years after the ratification of the Treaty. 'I believe I may say without giving any one offence', wrote Defoe, 'that a firmer union of policy with less union of affection has hardly been known in the whole world. Nay, it will not, I believe, be offered or suggested that there is any visible increase of good will, charity, or love of neighbourhood between the nations since finishing the transaction of the late Treaty of Union.'[133] Surely James VI and I could never have considered a union marked by such sentiment as the realization of his dreams.

The question remains, however, whether the institutional and economic union established in 1707 was eventually able to reduce the tensions that prevailed between the two nations at the time of the Treaty, encourage the assimilation of the people of both countries and create a genuine British national consciousness. Concerning the persistence of national animosities it is difficult to obtain anything but fragmentary evidence. It appears, however, that at least until 1800, when the United Kingdom was extended to include Ireland as well as Great Britain and the question of British nationality became even more complex, Englishmen and Scots did not view each other in ways that encouraged sentiments of national solidarity. When English MPs commented on the odd appearance and incomprehensible dialect of their Scottish

[131] Daiches, *Scotland and the Union*, 143–4; B. Hall, 'Daniel Defoe and Scotland', in *Reformation, Conformity and Dissent*, ed. R. B. Knox (London, 1977), 228–9.
[132] W. S. McKechnie, 'The Constitutional Necessity for the Union of 1707', *SHR* 5 (1908), 52. This was most evident in the *Worcester* incident of 1706. See Daiches, *Scotland and the Union*, 188.
[133] [D. Defoe], *Union and No Union* (London, 1713), 4.

colleagues; when a Scot like William Tod was jeered by a London mob and told to 'get home you crowdie and be d——d to you'; when James Boswell felt his Scottish blood boil as an audience at Covent Garden shouted 'No Scots! No Scots! Out with them!' on the appearance of two Highland officers in 1764; when David Hume could report that all Englishmen hated him because he was a Scotsman; and when John Wilkes was able to stir up 'rage against the Scots' during the anti-Bute agitation of the 1760s, the goal of creating a united British people could hardly be said to have been realized.[134] As one historian observed in 1799, 'a long period of years, passed in the mutual intercourse of peace and friendship, has scarcely worn out the traces of national antipathies in north and south Britons'.[135] In 1818 a historian of Glasgow still referred to the 'detested treaty' of 1707.[136]

Even if these strong national antipathies had abated, as they eventually did in the nineteenth and twentieth centuries,[137] their reduction or disappearance by itself could not have blended the English and Scottish peoples into one nation. In order for that to come about, it was necessary for a certain amount of assimilation of one people to the other to take place and for the participants in that process to develop a distinctly British national consciousness. This process appears to have begun in the eighteenth century, when large numbers of Scots, attracted by the cosmopolitan and both politically and commercially active world of London, moved south to seek fame and fortune in so doing became assimilated to the dominant English culture, language, and way of life. These men included David Hume, Alexander Carlyle, James and Robert Adam, Bishop Burnet, William Murray, the literary entrepreneur David Mallet, and James Thomson, author of the lyrics to 'Rule Britannia', not to mention Lord Bute and a

[134] N. T. Phillipson, 'The Scottish Whigs and the Reform of the Court of Session, 1785–1830', Ph.D. thesis (Cambridge, 1967), 64–5; *Boswell's London Journal, 1762–1763*, ed. F. A. Pottle (London, 1950), 71–2. *The Letters of David Hume*, ed. J. Y. T. Greig (Oxford, 1932), i. 470; J. A. Smith, 'Some Eighteenth-Century Ideas of Scotland', in *Scotland in the Age of Improvement*, ed. Phillipson and Mitchison, 109.

[135] E. Marshal, *The History of the Union of Scotland and England* (Edinburgh, 1799), 12.

[136] J. Cleland, *Annals of Glasgow* (1818), ii. 61.

[137] These antipathies did not abate completely. See 'An Overseas Englishman', *England* (London, 1922), 8–9.

large group of Scottish statesmen who situated themselves at Court.[138] All these men supported the union and considered themselves citizens of Great Britain. Their national identity, however, remained distinctly Scottish. As Janet Adam Smith has observed, 'consciousness of being British is rather for special occasions'; in their everyday consciousness, these men thought of themselves as Scots.[139] They accepted Britain as their state, their political society, and the focus of their civic consciousness, but they defined their national identity as Scottish or, more ambiguously, North British. They exemplified a new eighteenth-century phenomenon that has persisted until the present day: Scottish patriotism within the union, a sentiment that the Treaty of Union, with its exclusively civic, political orientation, had made possible.[140]

Another illustration of this mentality can be found among the landed gentlemen of eighteenth-century Scotland who later became known as the improvers. Although these men often remained in Scotland, they none the less manifested many of the Anglophilic, assimilationist, 'outward-looking', tendencies of Scots like Hume, Carlyle, and the Adams. Recognizing that they as Scots had not yet received all the privileges of the English landed class and that the Scottish economy was in a backward state, they set out to 'complete the union', to make themselves and the Scotland in which they had considerable power equal partners in the union. This meant, in practical terms, obtaining all of the privileges of the English gentry *vis-à-vis* their inferiors, exploiting English and other non-Scottish techniques to make the Scottish economy a viable counterpart to the English, and importing into Scotland English political and legal institutions, such as civil jury trials and the militia. The improvers, therefore, became, like the more cosmopolitan Scots who migrated south, 'super first-class citizens of Great Britain'.[141] At the same time,

---

[138] J. Carswell, *From Revolution to Revolution: England 1688–1776* (New York, 1973), 61, 105; Smith, 'Eighteenth-Century Ideas of Scotland', 107–1; Harvie, *Scotland and Nationalism*, 17.   [139] Smith, 'Eighteenth-Century Ideas', 108.

[140] See Harvie, *Scotland and Nationalism*, 16, 61; T. E. S. Clarke and H. C. Foxcroft, *A Life of Gilbert Burnet, Bishop of Salisbury* (London, 1907), 480–1.

[141] R. Mitchison, 'Patriotism and National Identity in Eighteenth-Century Scotland', in *Nationality and the Pursuit of National Independence*, ed. T. W. Moody (Belfast, 1978), 94.

however, as the beneficiaries of Scottish as well as English society, they retained a distinct identity as Scottish, at times even priding themselves as morally and religiously superior to their English counterparts. The improvers, therefore, exhibited another form of Scottish nationalism within the union, in its own ambivalent way a testimony to the fact that national consciousness after 1707 did not assume a predominantly British character.

Among other segments of Scottish society, less enamoured of the union, Scottish national consciousness expressed itself more forcefully. The Jacobites, who sought to undo the union, represented an extreme position, but they were not alone. Numerous presbyterians continued their opposition to the Treaty, viewing the kirk as the soul of the nation. Poets like Robert Fergusson, who protested against the unequal partnership of the union, continued to use the old Scottish language.[142] Lower-class Scots, who had little hope of exploiting the union for their benefit, continued to consider themselves Scottish in every way. Finally, the Highlanders must be considered as a special case. The strength of Jacobitism naturally turned them against the union, but they identified more with the culture and geography of the Highlands than with Scotland as a nation.[143]

Once we leave the eighteenth century the question of separate English and Scottish national identities becomes more problematic. On the one hand, the technical criterion for citizenship after 1800 was membership in an expanded United Kingdom, a state that included all of Ireland until 1922 and the province of Northern Ireland since then. The incorporation of Ireland into a larger, no longer exclusively British state, coupled with the emergence of Irish, Welsh, and Scottish nationalist movements, has made it more difficult for the state to serve as the focus of national loyalties. The United Kingdom is today, as Richard Rose has clearly demonstrated, a multi-national state.[144] On the other hand, during the

---

[142] Smith, 'Eighteenth-Century Ideas', 117.

[143] Hechter, *Internal Colonialism*, 112–13 *et passim*.

[144] Rose, 'United Kingdom as a Multi-National State'. See also A. Reid, 'Digging up Scotland', *New Yorker*, 5 Oct. 1981, 69.

nineteenth and twentieth centuries England and Scotland, having jointly engaged in both imperial and industrial development, fought together in numerous international conflicts, witnessed a reduction in the differences between their legal systems, and experienced a large degree of inter-marriage and geographic mobility between them, have achieved among both Scots and Englishmen, as well as the Welsh, a sense that they are all members of something loosely termed 'British society'.[145] That society, moreover, cannot be thought of in strictly political terms. Arthur Balfour was not in error when he claimed that over the centuries the Scots and English had been welded together 'in an inseparable unit'.[146] Nor is the political scientist Karl Deutsch wrong when he refers to Britain as a national state encompassing Englishmen, Welshmen, and Scotsmen who still retain a certain measure of 'ethnic, linguistic and group diversity'.[147] There is no doubt that these three peoples share enough cultural similarities to make being British mean more than simply complying with the dictates of a British state and assisting that state when it is in need. In some sense a British nation does exist.

In an attempt to reconcile these apparently contradictory views of contemporary British nationalism, one can refer to a secondary British national consciousness, a sentiment which even the people of Northern Ireland share. This secondary British nationalism, which usually receives its most forceful expression in times of emergency, is perfectly compatible with a primary English, Scottish, or Welsh national consciousness. It invites comparison with the outlook of those eighteenth-century Scots who on certain occasions defined themselves as British without in any way denying their basic Scottish nationality. The primary reference points for national consciousness in Britain today, just as in the eighteenth century, remain the ancient national communities, not their more recent composite.

[145] D. Snowman, *Britain and America: An Interpretation of their Culture, 1945–1975* (New York, 1977), 29.
[146] W. R. Scott, 'The Fiscal Policy of Scotland before the Union', *SHR* 1 (1904), 173.
[147] K. W. Deutsch, 'Nation-Building and National Development', in *Nation-Building*, 7.

V

If in fact Scottish national consciousness has persisted for nearly three centuries after the union, the question remains what factors have sustained it. Nations, as defined above, are self-conscious communities, sharing some common cultural or ethnic characteristics, which have the desire to express their identity in the form of distinctive political institutions. Traditionally the institutions of the state, either in their actual or anticipated form, have provided the opportunity for this type of national expression. Ever since the union, however, Scotland has been deprived of its own autonomous political institutions, and the desire to create such institutions, which is the goal of the Scottish nationalist movement, is not shared by even a majority of the Scottish people. Because of Scotland's failure to possess these institutions, Marx and Engels refused to recognize Scotland's status as a nation.[148]

Scotland, however, has possessed ecclesiastical and judicial institutions that have filled the political gap created in 1707. The first of these, the General Assembly of the national church, had for a long time prior to the union performed valuable political functions in Scottish society and had in fact achieved much more success than the Scottish parliament in attracting national loyalties. After the union the General Assembly declined in influence, mainly because of the increasing secularization of the Scottish aristocracy, but it still represented the nation in a real sense.[149] Much more important than the General Assembly were the national courts of the country. Guaranteed their independence from English judicial authorities (except the House of Lords) by the Treaty of Union, the courts stood as symbols of the distinctness of Scottish public life and even held their sessions in the old Parliament House. The men who served as judges represented the highest autonomous political authority within the country, and the power they exercised was national in scope.[150]

In addition to the General Assembly and the courts,

---

[148] Harvie, *Scotland and Nationalism*, 25.

[149] On the decline of the church in this regard see A. L. Drummond and J. Bullock, *The Scottish Church, 1688–1843: The Age of the Moderates* (Edinburgh, 1973), 177.

[150] D. Daiches, *The Paradox of Scottish Culture: The Eighteenth-Century Experience* (London, 1964), 36–67.

Scotland has acquired a number of administrative institutions which, while deriving their authority from a centralized British parliament, exercise power in Scotland on a virtually autonomous basis. In this regard the historic tendency of the British central administration to delegate its authority to county and regional bodies has worked to Scotland's advantage. In almost every area of public administration Scotland has its own boards and departments. These institutions, taken together, have helped to maintain Scotland as a semi-independent polity and have given the Scottish people additional sources of national consciousness.

One can argue, therefore, that Scottish national identity has received support from a sufficiently broad range of 'distinctive' political institutions, even if those institutions have not been those of a sovereign Scottish state. None of these institutions has the capacity to foster national sentiment in the way that a Scottish parliament or even a devolved Scottish assembly could, but collectively they have encouraged Scots to recognize that they constitute a distinct community within a united Britain.

Scottish national consciousness has also gained strength from additional, more properly social and cultural sources. A set of distinctly Scottish educational institutions, for example, has served the purpose of maintaining Scottish culture and preventing the inevitable assimilation that a fully integrated British system would have brought about. The foundation for a self-contained Scottish educational system was laid in 1616, when the first efforts were made to establish a school in every parish, an experiment in national education that had no parallel in England or continental Europe. This system, imperfect though it may have been, was improved in 1696, remained intact after the Union, and continues to the present day to operate on an almost autonomous basis.[151] In the upper levels of education a parallel development occurred. In the eighteenth century Scotland had four universities, whereas England had only two, and although some students from each country attended the universities of the other, the ample and in many respects more promising educational opportunities in

[151] R. Pares, 'A Quarter of a Millenium of Anglo-Scottish Union', *History*, 39 (1954), 237, 239.

Scotland prevented the English universities from becoming agents of assimilation. Another Scottish educational and cultural institution, the National Library of Scotland, itself the child of another distinctly Scottish institution, the Faculty of Advocates, has continued to exist independently of the British Museum and, more recently, the British Library.

The persistence of a distinct Scottish history, today symbolized by the continued existence of separate faculties of Scottish history at some Scottish universities, has also made an important contribution to the survival of the Scottish nation. The recognition of a common past, in which the ethnic, cultural, and political foundations of nationhood were estab-lished, has usually been an essential ingredient in the process of nation-building. In the sixteenth and seventeenth centuries it was uncertain whether Scots would have this type of national history. Scottish scholars at this time lacked many of the sources, especially the legal records, from which that history could be written.[152] Those who did venture into this murky area usually adopted a non-nationalist perspective, choosing to view Scotland in one way or another as part of a broader British or European world. It appeared that if there was to be a Scottish history, it would be part of a larger British history, such as the one Bacon planned to write.[153] Bacon's project, however, was stillborn, and nothing resembling a British history, except histories of England writ large, appeared in the seventeenth, eighteenth, and nineteenth centuries. This meant that Scottish history could establish itself as a separate area of enquiry. This history was not anti-unionist, and therefore it did not lead to the development of a full-blown Scottish nationalism. But it did provide support for the widespread belief that the Scottish people constituted a distinct nation.

The continued existence of the Scottish nation, like the continued independence of the Scottish church and legal system, stands as a lasting commentary on the Treaty of Union of 1707. The men who negotiated the Treaty had no interest in creating a united British nation and therefore

---

[152] J. G. A. Pocock, 'British History: A Plea for a New Subject', *Journal of Modern History*, 47 (1975), 601–24.

[153] *L&L* iii. 249–52. See also Craig, *De Unione*, 468.

enabled the Scots to preserve their own national identity within the union. They thus created a curious structure, a unitary state that was multi-national. This structure differed dramatically not only from the Scottish and English states that it integrated but also from the type of united British community envisaged by James VI and I and by most of the unionists of the early seventeenth century.

7

Conclusion: The End of Regal Union

THE Treaty of Union of 1707 concluded a long chapter in the history of Anglo-Scottish relations. It marked the end of the regal or personal union which had joined the two countries, with only one brief hiatus, since 1603. Instead of merely sharing the same king, the two kingdoms were now joined in one body politic. After 1707 Englishmen and Scotsmen elected representatives to the same parliament, paid the same taxes and customs, competed for the same governmental posts, joined the same trading companies, and found themselves subject to many of the same governmental authorities. The new union was also a more permanent arrangement. Whereas the personal union would continue only as long as both countries followed the same order of succession to the throne, the incorporating union guaranteed that the laws of succession in both kingdoms would always be identical.

Since the incorporating union has survived, even to the present day, it is tempting to see it as the product of inexorable historical forces. Like those early eighteenth-century unionists who saw the hand of Providence directing England and Scotland towards perfect union, historians have often expressed the belief that a 'benignant fate' steered England and Scotland from the Union of the Crowns to the Union of the Parliaments.[1] Even if they do not adopt an explicitly providentialist or determinist point of view, they tend to see the advocates of incorporating union either as men of vision or at least as men who were ahead of their time. Thus James VI and I and Oliver Cromwell, whatever their drawbacks, are

---

[1] G. Dent, *A Thanksgiving Sermon Preach'd on the First Day of May, 1707* (London, 1707), 12; [Grant], *The Patriot Resolved*, 5; Nicolson, *Leges Marchiarum*, p. i; R. S. Rait, *An Outline of the Relations between England and Scotland (500–1707)* (London, 1901), 146.

given credit for anticipating the union that eventually was implemented. Unlike their myopic contemporaries, *they* knew that personal union was both artificial and impermanent and tried to implement a more ·durable, even if less popular, alternative.

The providentialist approach to the union question has two main weaknesses. The first is that it takes early unionists out of context and makes them appear to be much more modern or forward-looking than in fact they really were. Nothing can be more dangerously ahistorical than to view Oliver Cromwell as a prophet of 1707, even if many aspects of his union programme were adopted in the Treaty of that year. Union meant something quite different to Oliver Cromwell from what it did to Queen Anne, and many of the measures that he took to implement union would have been anathema to unionist and anti-unionist alike in 1707.

The second weakness of the providentialist approach is that it writes off the regal union of 1603 as a pre-ordained failure. That it ultimately proved to be unworkable cannot be denied, but one cannot conclude from this outcome that it was destined to fail. A number of dynastic unions, most especially those arranged by Habsburg kings, did in fact survive the seventeenth and even the eighteenth century,[2] and there was just as much cause for optimism in 1603 that the regal union of England and Scotland would survive in its strictly dynastic form than if the two states were to be joined in one body politic. Indeed, from the vantage-point of 1603, there was more hope for a regal than for an incorporating union, since the latter was without precedent. In no other European country had two independent kingdoms been joined in one state except as a result of conquest. For someone who took his

[2] The main step in the transformation of the various provinces and kingdoms of Spain into a centralized Spanish state was the publication of the Nueva Planta in 1716, but even this reform did not create an inseparable monarchy such as was established by the Anglo-Scottish Treaty of 1707. As late as 1835 there was no guarantee that the crowns of Castile and Aragon would follow the same line of succession. Elliott, *Imperial Spain*, 377; J. N. H. Hilgarth, *The Spanish Kingdoms, 1250–1516*, Vol. ii: *1410–1516* (Oxford, 1978), 614. The union of Norway and Sweden, which was formed in 1815, was also a personal union in which the main link between the two countries was their common king. See R. E. Lindgren, *Norway–Sweden: Union, Disunion, and Scandinavian Integration* (Princeton, NJ, 1959), 26–7.

historical precedents seriously, the union project of James VI and I was a dangerous novelty.

Whatever chances regal union had in 1603, it did not survive. By the beginning of the eighteenth century there was virtually unanimous agreement that the Union of the Crowns could not be continued in its present form and must give way to incorporation, federalism, or complete independence.[3] Why did this situation arise? Once we abandon the explanation that the regal union was destined to fail, we can isolate three closely related reasons for the breakdown: diplomatic conflict, constitutional change, and the subordination of Scottish to English interests.

In any analysis of the change that took place in Anglo-Scottish relations in the early eighteenth century, disagreement over the conduct of foreign policy must play a prominent part. More than any one factor, the threat made in 1703 that Scotland would refuse to participate in a future 'British' war forced England to recognize that the regal union could no longer ensure the protection of its interests and that therefore it should begin serious pursuit of an incorporating union.

At the time of the Union of Crowns it did not appear that a diplomatic conflict of this sort would ever arise. As the authors of many early seventeenth-century union tracts noted, James by his mere accession to the English crown had eliminated the possibility of an 'inland enemy' and ended once and for all the 'Auld Alliance' between Scotland and France, which had always been the 'bridle of England'.[4] No wonder that Henri IV had considered supporting a Catholic Habsburg candidate for the English throne in order to prevent the union.[5] Once the

---

[3] On the failure of the regal union see [Buchan], *Memorial*, 6; Seton, *Interest of Scotland in Three Essays*, 110; *Reflections upon a Late Speech by the Lord Haversham in so far as it relates to the Affairs of Scotland* (Edinburgh, 1704), 22–3; [Paterson], *An Inquiry into the Reasonableness and Consequences of an Union with Scotland* (London, 1706), 81–3; Pryde, *Treaty of Union*, 13; Dicey and Rait, *Thoughts on the Union*, 137, 141, 177–9.

[4] A. J. Loomie, SJ, 'Philip III and the Stuart Succession in England', *Revue Belge de Philologie et d'Histoire*, 43 (1965), 510. According to Lord Belhaven, King Henri IV had done all that he could to prevent the union from being strengthened. J. Hamilton, 2nd Lord Belhaven, *A Speech in Parliament on the 10 Day of January, 1701* (Edinburgh, 1701), 5.

[5] M. Lee, Jr., *James I and Henri IV: An Essay in English Foreign Policy* (Urbana, Ill., 1970), 10–11.

union was accomplished, it appeared that England and Scotland would always have the same foreign policy.

Nevertheless, the regal union did not remove the threat of independent Scottish activity in foreign affairs. As one Italian observed shortly after 1603, 'England hath gotten no great catch by the addition of Scotland; she had only got a wolf by the ears, who must be held very fast, else he will run away to France'.[6] Until the two countries were fully united, the possibility always existed that Scotland might refuse to support an 'English' war effort. During the war against Spain and France in the late 1620s, for example, the king encountered great difficulty raising troops in Scotland, especially after military activity was expanded to include France in 1627.[7] These problems in Scotland help to explain why the English Secretary of State, Sir John Coke, proposed at this time that England, Scotland, and Ireland be joined 'in a strict union and obligation each to other for their mutual defence', an arrangement modelled on a recent association of the Spanish kingdoms for similar purposes.[8] Another indication of Scotland's independence in foreign affairs was its decision to negotiate directly not only with the Netherlands but also with France during the 1640s.[9] These efforts foreshadowed the course of action threatened by the Scottish parliament in 1703.

The diplomatic conflict that developed in the early eighteenth century could not have arisen unless the consitutions of both England and Scotland had undergone significant changes since 1603. At the time of the Union of the Crowns the foreign policy of both England and Scotland came under the direct and immediate control of the king himself, who decided matters of war and peace by virtue of his prerogative. So long as this situation continued, there was little likelihood that Scotland would have been able to take the steps it did in

---

[6] *Bella Scot-Anglica*, 19. On the continuation of the Auld Alliance after 1603 see Smith, *Studies Critical and Comparative*, 30; 'Memoirs Concerning the Ancient Alliance between the French and the Scots and the Privileges of the Scots in France', *Miscellanea Scotica*, 4 (1820), 47–9.

[7] Lee, *Road to Revolution*, 79–82.

[8] *CSPD Addenda, 1625–1649*, 241–2.

[9] *Hamilton Papers*, 235; *Correspondence of the Scots Commissioners*, p. xxv.

1703.[10] It might have resisted the king's efforts to raise troops, as it had in the 1620s, but its parliament could not have directly challenged the foreign policy of the king. In the late seventeenth century, however, the king lost much of his control over foreign policy in both kingdoms. In England parliament tightened its control over the purse, thereby acquiring much leverage in determining foreign policy, while in Scotland the main mechanism that the king had used to gain parliamentary approval of his policies, the Committee of Articles, was abolished in 1690. These changes meant that the king might be pressured to follow two different foreign policies. Theoretically he might even be forced to declare war on himself. In practical terms the changes meant that in the event of a conflict between the English and Scottish parliaments, the policy of the more powerful English parliament would prevail, as it did in 1702.[11] By the same token, however, Scotland could now exercise an effective veto over its participation in a 'British' war effort or simply choose another king, the two courses of action it threatened to take by passing the Security Act and the Act anent Peace and War in 1703. That legislation was in a very real sense the product of the constitutional revolution of the seventeenth century.

But why did Scotland feel that it needed to threaten dynastic divisions and an independent foreign policy in the first place? In a certain sense the diplomatic and constitutional reasons for the failure of the personal union were subordinate to the main one, which was the growing awareness in Scotland that England was using the regal union to govern Scotland in England's interest and to bring their country into a 'servile dependence' upon the southern kingdom.[12]

During the reign of James VI and I, Scots had little reason to view the regal union in this way. To be sure, James did use his position in England to implement unpopular changes in Scotland, especially in ecclesiastical matters. He can also be held responsible for depriving Scotland of its own court as a

---

[10] As Lord Ellesmere asked in his speech on the *post-nati*, 'Can there be wars between the King of England and the King of Scotland . . . so long as there is but one King?' Knafla, *Law and Politics in Jacobean England*, 246.

[11] McKechnie, 'Constitutional Necessity for the Union', 55.

[12] [J. Sage], *Some Remarks on the Late Letter from a Gentleman in the City* (1703), 10.

result of his decision to rule both his kingdoms from Westminster. But James cannot reasonably be accused of adopting policies that were inherently hostile to Scottish interests.[13] Indeed, James, who frequently found himself defending his native land against English verbal assaults, was much more vulnerable to the charge of ruling England in Scotland's interest than of ruling Scotland in England's.

With Charles I the recognition of the debilitating effects of regal union slowly began to become apparent. In Charles the Scots acquired a king who had none of his father's affection for the northern kingdom and who certainly did nothing to promote Scots at court. When he tried to open the North Sea fisheries to Englishmen, Scotsmen got their first taste of the type of economic exploitation by England that later became systematic. And when Charles tried to implement his ecclesiastical policies in the 1630s there was ample reason to fear the eventual destruction of Scottish independence. As David Stevenson has observed, 'the plight of the kirk came to be taken as symbolic of Scotland's position under the union of the crowns', and the revolution that began in 1637 was to a great extent a rejection of the alien, anglicizing influences that followed in its train.[14]

With the death of Charles I and the establishment of the Republic, the regal union was of course interrupted, and it was replaced briefly by Cromwell's incorporating union. Although Scotland was ruled throughout this period as if it were a conquered country, it is difficult to claim that its interests were regularly subordinated to those of England. Like most incorporating unions, the Cromwellian experiment did not allow one region within the state to practise a policy of systematic discrimination against another. Indeed, some of the main complaints about the economic aspects of the Cromwellian Union came from English merchants who felt that English mercantile interests were being ignored.[15]

The contrast between Scotland's commercial position

---

[13] Defoe, *Essay at Removing National Prejudice*, Pt. I, 7–8, claims that James suppressed the 'true interest' of Scotland, but that was because he failed to bring about an 'entire union', which Defoe believed James could have done.

[14] Stevenson, *Scottish Revolution*, 313, 319.

[15] *Diary of Burton*, iv. 57.

during the Cromwellian period and after the Restoration, when the regal union was restored, made the dangers inherent in the Union of the Crowns more apparent than they had been at the end of the reign of Charles I. Shortly after Charles II came to the throne the English parliament passed a number of statutes discriminating against Anglo-Scottish trade. These measures exposed one of the main weaknesses of the personal union. Ever since the early seventeenth century the *post-nati* of either country had been considered naturalized in the other country. Their naturalization was based on the personal bond of allegiance that had existed between them and their king, who ever since 1603 had been ruler of both England and Scotland. According to the judges who had determined this point of law, however, the union and the naturalization that followed from it had nothing to do with the laws of both countries, which remained distinct in 1603. There was nothing, therefore, except an unlikely royal veto, to prevent the parliament of either country passing laws that treated the subjects of one country differently from those of the other. Thus the absurd situation arose in the 1660s that men who were naturalized citizens in England could be treated as if they were aliens, a situation that the English Alien Act of 1705 threatened to make official.[16] This act offered legislative proof that the regal union of 1603 had failed.

As the seventeenth century progressed, and as the king came more under the control of his English parliament, Scotland found the regal union even more intolerable. The worst blow came in the late 1690s, when the English parliament persuaded William III to sabotage the efforts of the Scottish parliament to establish an overseas colony at Darien. Even if the colony had little chance of success to begin with, its failure served as the clearest indication that England could use the regal union to protect English economic interests in the face of Scottish competition. The episode also exposed the consitutional weakness of the regal union, since William, as king of England, had taken steps to undermine a policy that he himself had previously approved as king of Scotland.[17]

---

[16] For a constitutional defence of this policy see Seafield's 'Memorial', HMC, *Laing MSS*, ii. 127.

[17] Dicey and Rait, *Thoughts on the Scottish Union*, 146–51.

In addition to making the Scots only too painfully aware of the weakness of the regal union, the Darien episode convinced the king himself that the union could not be maintained in its present form. One of the reasons he had agreed to withdraw his original support for the Scottish venture was that it had angered Spain, which claimed sovereignty over the Isthmus of Darien. Indeed, Scottish attempts at settling the area had actually led to a brief armed conflict with Spain in 1700. Scottish economic enterprise, therefore, threatened to work against the foreign policy which the king was pursuing and which his English parliament was supporting. Because of this conflict William realized that he could no longer 'drive two unlinked horses', and in 1700 he took the initiative in proposing negotiations for a union.[18]

Once it became readily apparent that the personal union could not survive, the question of what would take its place arose. Complete independence was of course a possibility, but it was more of a threat than a viable option. There was little serious support for it in either kingdom, and the strength of English opposition to it would have led to war, if necessary, to prevent it. The real choices were some sort of federalism (or, more properly speaking, confederation) and incorporation. The federalist position had great popular appeal, mainly because it allowed for the fullest expression of Scottish patriotism, and the case for it has been revived in discussions of devolution in the twentieth century. But in the early eighteenth century federalism, in addition to being completely unacceptable to the English, offered very few guarantees that Scottish interests would be protected. The main drawback from the Scottish viewpoint was that the federation would consist of two partners of unequal strength, a situation that would inevitably lead to the domination of one country by the other, the very situation that had made the regal union intolerable. Successful federations, at least during the last three centuries, have been those that have comprised many political units, an arrangement that prevents the stronger from dominating the weaker.[19]

[18] Mitchison, *Lordship to Patronage*, 125.
[19] McKechnie, 'Constitutional Necessity for the Union', 61–2; Chamberlen, *Great Advantages*, 8.

The men who negotiated and approved the Treaty of Union of 1707 rejected federalism and adopted a plan for incorporating union, but that union, as this book has emphasized, was not complete. The two parliaments were reduced to one, and Scots and Englishmen were given complete freedom of trade in either kingdom, but the administration of the two countries was never fully unified, and their laws and churches were kept separate. This arrangement, which lacks any parallel in Europe, has been described as quasi-federalism. Whatever one may call it, it has been responsible for maintaining Scotland as a 'satellite' of England.[20] If the union of 1707 had been complete, if Scotland had been fully incorporated into England, satellite status would never have become a possibility.

The continuation of Scotland as a satellite after 1707 suggests that the Treaty of that year did not completely solve the problem that had caused the failure of the personal union. It was still possible after 1707 for a predominantly English parliament and a predominantly English ministry to make decisions that were not in Scotland's interest. If the union had been complete, it would have become increasingly difficult to identify such interests, and in the twentieth century it would have been much more difficult for a Scottish nationalist movement to arise. In a certain sense, the Treaty of Union, by preserving a Scottish nation, has either created or perpetuated forces that have worked towards its modification or repeal. The Treaty of Union may claim the status of fundamental law in Britain, but the actual union that it achieved is much more fragile than it appears.

The Treaty of 1707 brought an end to the regal union of 1603, but it did not mark the end of such unions in British history. For the remainder of the eighteenth century a regal bond continued to tie Ireland, the king's third kingdom, to Britain, and a similar one linked Britain to its American colonies. There were of course significant differences between Scotland and these territories. Neither Ireland nor any of the American colonies, for example, ever possessed Scotland's sovereign status. Having been either conquered or settled by

---

[20] Cowan, 'Union of the Crowns', 122.

Englishmen, they acquired English law and a highly disputed subjection to the English parliament. From the very beginning, therefore, there was a closer connection between these territories and the English state than there was between England and Scotland in the seventeenth century.[21]

Nevertheless, Ireland and the colonies did, like Scotland, acquire representative assemblies of their own, and the nationality of their people was determined at least in part by *Calvin's Case*. In the late eighteenth century, moreover, the relationship between Britain and these territories had to be redefined, just as the relationship between England and Scotland required redefinition in the early eighteenth century. Both in Ireland and America, just as in Scotland, regal union came under attack, mainly because the government at Westminster tried to rule these areas in Britain's interest. In Ireland this conflict had an outcome similar to that of 1707.[22] In 1800, after a long debate regarding the relative merits of Irish independence and closer union, Ireland was joined to Britain in an ill-fated incorporating union that lasted only until 1922. In America the result was quite different. Although various plans for legislative union with Britain were considered,[23] thirteen of the colonies proclaimed their independence in 1776 and secured it militarily in 1783.

Parallels between Scotland and Britain's overseas possessions might strike us as inappropriate today. In the early seventeenth century, however, it appeared that the union with Scotland was the first step in the establishment of the new English empire that Elizabethan writers had anticipated.[24] It

---

[21] On the differences between the Scottish and both the Irish and American situations see McIlwain, *American Revolution*, 37, 78–80; *Hibernica* Pt. II, 112; Craw v. Ramsey in *The Reports and Arguments of . . . Sir John Vaughan* (London, 1706), 278. For the American argument that the union with Britain was strictly regal, see Kettner, *American Citizenship*, 158–60.

[22] The Union with Ireland Act, however, was a unilateral act of the British parliament, similar to the English Act of Union with Wales. There was no treaty and no corresponding act of the Irish parliament, as there had been in with Scotland in 1707. See R. K. Murray, 'The Anglo-Scottish Union', *Scots Law Times*, 4 Nov. 1961, 161–2.

[23] One of these plans called for American representation in the British parliament, 'as in the cases of Chester, Durham, Wales and Scotland'. J. P. Boyd, *Anglo-American Union: Joseph Galloway's Plan to Preserve the British Empire, 1774–88* (Philadelphia, 1941), 90–1.

[24] Parry, *The Golden Age Restor'd*, 16; Shakespeare's play *Cymbeline*, which was written in 1608 and which emphasizes the ideal of a united Britain, uses much

is true that King James refused to consider his native kingdom as an imperial possession. Repeatedly he told his parliaments and his advisers that he would not rule his northern kingdom as a conquered and slavish province. Nevertheless, the possibility of English imperialism always lay just beneath the surface during the seventeenth century,[25] and when England established garrisons in Scotland during the 1650s, it looked as if a truly imperial rule would be inaugurated. Of course it did not turn out that way, either in 1654 or in 1707. Instead of becoming an imperial possession Scotland became incorporated into the English state and thus became a part of the core state that lay at the centre of the eighteenth-century empire. The Scottish ruling class, moreover, took very well to their new imperial role and participated vigorously in the settlement and governance of Britain's overseas possessions, which were now just as much theirs as England's.[26] After 1707 personal union was an arrangement that linked them not to England but to *their* overseas possessions.

One of the most vivid illustrations of the nature of the personal or regal union that ended in 1707 is the panel depicting the union on the ceiling of the Banqueting House at Whitehall. This painting, which Charles I commissioned Rubens to do in honour of King James, shows the king, seated on his throne, pointing to two women, who represent England and Scotland, and a new-born baby, who symbolizes a united Britain. Each woman holds a crown over the child's head while a goddess, most likely Minerva, joins the two crowns.[27] The painting stands as the perfect expression of the union that James accomplished. It was essentially a regal union, symbolized by the merging of the two crowns, and a personal union, made in his blood. Its personal character is symbolized

imperial imagery, including the Roman eagle winging its way westward to vanish into the British sun. L. S. Marcus, *Shakespeare and the Unease of Topicality* (Berkeley, Calif., forthcoming). On the Elizabethan idea of empire see generally Yates, *Astraea*, 38–59.

[25] In 1638, for example, the Earl of Strafford proposed that England impose its law and ecclesiastical government on Scotland and govern it 'by the King and Council of England in a great part', just as in Ireland, R. Bagwell, *Ireland under the Stuarts and during the Interregnum* (London, 1909), i. 237. See also the speech of Sir William Temple in the Parliament of 1659. *Diary of Burton*, iv. 132.

[26] A. Calder, *Revolutionary Empire* (New York, 1981), 426–7, 448–9, 474–5.

[27] D. J. Gordon, 'Rubens and the Whitehall Ceiling', in *The Renaissance Imagination*, ed. S. Orgel, 38–41.

by both the two women (who recall James's marital analogies and the references to his two wives) and the child, who personifies the new nation. The painting is also, like the Jacobean union itself, regno-centric. All eyes, including those of Minerva, who represents James's wisdom, are turned towards the king, who directs the entire operation. There is much historical truth in their glances. The early seventeenth-century union project was very much James's personal plan, and the limited success that it achieved was due to his, rather than his parliaments' activities.

There is no painted ceiling depicting Queen Anne and the Union of 1707.[28] By that time, of course, the cult of king-worship that prevailed in the sixteenth and early seventeenth centuries had long since died out, so if a ceiling painting had been commissioned, we would not expect to see Anne depicted in such an omnipotent or even such a sagacious manner as James. We also would not expect to see a picture in which all eyes were turned towards Anne. The queen was not completely passive in the union negotiations. She used her prerogative powers to name the commissioners from both kingdoms, and as queen of both countries she technically concluded the Treaty with herself. But the main impulse to union in 1707 did not come from her; here the contrast with 1604 is clear. The Union Treaty was the result of the English government's recognition that incorporating union was necessary for England's security and the Scottish government's decision, for many different reasons, to co-operate with them. Finally, if there had been a ceiling painting of Anne, we would not expect to see the crowns and the young child. The Union of 1707 was not essentially a regal and personal union, as James's union had been. It did, of course, guarantee that there would now be one inseparable British monarchy, and it permanently united the two crowns, but its main achievement was the union of the parliaments and the creation of a united British state.

[28] In St Stephen's Hall there is a painting of the Queen receiving the Articles of Union from the English and Scottish commissioners, but the work was done in the 20th century.

Bibliography

MANUSCRIPTS

Beaulieu Palace House

Papers on Scotch Affairs. 'The Devine Providence in the Misticall and Real Union of England and Scotland'.

Bodleian Library, Oxford

Calendar of Carte MSS

Carte MSS 74, 82, 84, 85, 86, 105.

Clarendon MS 133. Notes on the King of England's sovereignty over Scotland.

MS e Museo 55. A Treatise of Union by Sir Henry Savile (1604)

Smith MS 31. Proceedings against Mr. Hart of Magdalen College.

Tanner MS 75. Objections against changing the name of England.

Tanner MS 211. 'How England Should Have Homage and Fealty of Scotland'.

British Library, London

Add. MS 4158. Proposals concerning Judicature in Scotland.

Add. MS 12497. Gifts to Scottishmen 1603–1610.

Add. MS 24984. Petition of Scottish Commissioners, 1640.

Add. MS 32094. 'That the Crown of Scotland was not Subject to England'.

BM Loan 29/202. Harley Papers, 1582–1629.

Cotton MS Titus F IV. Papers and treatises regarding naturalization and trade with Scotland.

Egerton MS 1048. Instructions to Commissioners for Scotland, 5 Oct. 1659.

Harl. MS 158. Arguments of the masters of Trinity House against employment of Scottish ships in England.

Harl. MS 292. Treatises and Papers on the Union.

Harl. MS. 455. Articles of Anglo-Scottish Treaties, 1640–1.

Harl. MS 1305. 'A Discourse upon Marriage to be Made between the Three Kingdomes of France, Spain and Great Brittany' (*c.* 1606).

Harl. MS 1314. Parliamentary Objections against the Union, 1606.

Harl. MS 5220. John Doddridge's Treatise on the Prerogative.

Harl. MS 6842. Speech of Thomas Fuller in Parliament, 1606.

Harl. MS 6850. 'A Discourse on the Proposed Union' and other papers on the Union.

Royal MS 18A. 51. James Maxwell, 'Britaines Unioun in Love'.

Sloane MS 1786. 'Concerning the Communication of Laws'.

Stowe MS 187. Proposals of the Scottish Commissioners, 1640.

Stowe MS 158. 'A Briefe Replication to the Answers of the Objections against the Union' and 'Consideration on the State of Scotland in 1708'.

Edinburgh University Library

Laing MS I. 290. Proposed additions to the Act of Union, 1654.

Laing MS III. 249. David Hume, 'De Unione Britanniae . . . Tractatus Secundus.

Gonville and Caius College, Cambridge

MS 73/40. 'Pro Unione'.

Hampshire Record Office, Winchester

Jervoise MSS 44M69, no. 79. Justice Warburton's Collections.

Hatfield House, Hertfordshire

Salisbury MS 140. 102–3. William Bellenden's verses on the Union.

House of Lords Record Office, Westminster

Braye MS 50. Brief Collection of General Penal Statutes.

Huntington Library, San Marino, California

Ellesmere MS 1225. Heads of Union between England and Scotland.

Ellesmere MS 1226. Objections against the change of Royal Style.

Lincoln's Inn Library, London

Maynard MS 83. Journal of the Union Commissioners, 1604.

National Library of Scotland, Edinburgh

Advocates MS 31. 6. 12. David Hume, 'De Unione Britanniae . . . Tractatus Secundus'.

Advocates MS 31. 7. 7. George Mackenzie, 'Discourse concerning the Three Unions between Scotland and England'.

Wodrow MS Fol. lxxi. Proposals of the Scottish Commissioners, 1640.

MS 597. Lauderdale Papers.

MS 1019. Letter against the Union.

MS 2517. Designs for union flag.

Public Record Office, London

SP 14/5–10, 19, 22. 24, 26, 27, 28. Papers regarding the union, 1604–7.
SP 14/50/43. Letter on admission of Scots to English universities.
SP 14/57/104. Letter concerning David Hume's second union treatise.
SP 15/36/26–7. Papers regarding bill for recovery of small debts.
SP 16/131/27–9. Commissions for ecclesiastical causes, 1629.
SP 104/176. Proposals for the Commissioners of the Union, 1669.

Scottish Record Office, Edinburgh

CH 2/357/1. Kirk Session Records of Tranent.
PA 7/24. Parliamentary and State Papers, 1581–1651.

Trinity College Dublin

MS 635. William Clerk, 'Ancillans Synopsis'.

PRINTED PRIMARY SOURCES

a. Official Documents and Legal Records

Acts and Ordinances of the Interregnum, ed. C. H. Firth and R. S. Rait (3 vols.; London, 1911).
Acts of the Parliaments of Scotland, ed. T. Thomson and C. Innes (12 vols.; Edinburgh, 1814–44).
Calendar of State Papers, Domestic Series (London, 1856–72).
Calendar of State Papers Relating to English Affairs in Venetian Archives, ed. H. F. Brown *et al.* (London, 1864–1940).
Calendar of State Papers Relating to Ireland of the Reign of James I, 1603–1607 (London, 1872).
Cobbett's Complete Collection of State Trials and Proceedings for High Treason and Other Crimes, ed. W. Cobbett, T. B. Howell, *et al.* (34 vols.; London, 1809–28).
Constitutional Documents of the Puritan Revolution, ed. Samuel R. Gardiner (3rd edn., Oxford, 1906).
Constitutional Documents of the Reign of James I, ed. J. R. Tanner (Cambridge, 1961).
Correspondence of the Scots Commissioners in London, 1644–1646, ed. Henry W. Meikle (Edinburgh, 1917).
The Decisions of the English Judges during the Usurpation (Edinburgh, 1762).
Foedera, Conventiones, Literae et . . . Acta Publica, ed. Thomas Rymer (10 vols.; The Hague, 1739–45).
Journals of the House of Commons (London, 1803–).

Journals of the House of Lords (London, 1803–).

Register of the Privy Council of Scotland, ed. J. Hill and D. Masson (Edinburgh, 1877–1933).

Selected Justiciary Cases, ed. J. I. Smith (Stair Society, 27–8; Edinburgh, 1972–4).

Statutes of the Realm (12 vols.; London, 1810–28).

The Stuart Constitution, 1603–1688, ed. J. P. Kenyon (Cambridge, 1966).

Stuart Royal Proclamations, Vol. i: *Royal Proclamations of King James I, 1603–1625*, ed. James F. Larkin and Paul L. Hughes, (Oxford, 1973).

Vaughan, Sir John, *The Reports and Arguments of that Learned Judge, Sir John Vaughan* (London, 1706).

b. Letters, Memoirs, Diaries, and Papers

Baillie, Robert, *The Letters and Journals of Robert Baillie*, ed. David Laing (3 vols.; Bannatyne Club; Edinburgh, 1841–2).

Burnet, Gilbert, *Bishop Burnet's History of His Own Time* (London, 1838).

Boswell, James, *Boswell's London Journal, 1762–1763*, ed. F. A. Pottle (London, 1950).

The Clarke Papers, ed. C. H. Firth (4 vols.; Camden Society, 49, 54, 61, 62; London, 1891–1901).

Correspondence of King James VI of Scotland with Sir Robert Cecil and Others in England during the Reign of Queen Elizabeth, ed. John Bruce (Camden Society, 78; London, 1861).

The Cromwellian Union: Papers Relating to the Negotiations for an Incorporating Union between England and Scotland, 1651–1652, ed. C. S. Terry (SHS, 40; Edinburgh, 1902).

Diary of Thomas Burton, Esq., Member in the Parliaments of Oliver and Richard Cromwell from 1656 to 1659, ed. John T. Rutt (4 vols.; London, 1828).

The Earl of Stirling's Original Register of Royal Letters Relative to the Affairs of Scotland and Nova Scotia from 1615 to 1635, ed. Charles Rogers (2 vols.; Edinburgh, 1885).

The Egerton Papers, ed. Payne J. Collier (Camden Society, 12; London, 1840).

Goodman, Godfrey, *The Court of King James the First*, ed. J. S. Brewer (London, 1839).

The Hamilton Papers, ed. Samuel R. Gardiner (Camden Society, NS 27; London, 1880).

Historical Manuscripts Commission, *Hastings Manuscripts* (4 vols.; London, 1928–47).

Historical Manuscripts Commission, *Laing Manuscripts* (2 vols.; London, 1914–25).
——, *Mar and Kellie Manuscripts* (London, 1904).
——, *Portland Manuscripts* (10 vols.; London, 1891–1931).
——, *Salisbury Manuscripts* (23 vols.; London, 1895–1973).
——, *Third Report* (London, 1972).
Holles, Gervase, *Memorials of the Holles Family, 1493–1656*, ed. A. C. Wood (Camden Society, 3rd ser. 55; London 1937).
Hume, Sir David, of Crossrigg, *Diary of the Proceedings in the Parliament and Privy Council of Scotland, May 21, 1700–March 7, 1707* (Bannatyne Club; Edinburgh, 1828).
Hume, David, *The Letters of David Hume*, ed. J. Y. T. Greig (2 vols.; Oxford, 1932).
Illustrations of British History, ed. Edmund Lodge (3 vols.; London, 1791).
The Journal of Sir Simonds D'Ewes from the Beginning of the Long Parliament to the Opening of the Trial of the Earl of Strafford, ed. Wallace Notestein (New Haven, 1923).
The Lauderdale Papers, ed. Osmund Airy (Camden Society, NS 34; London, 1884).
The Letters and Life of Francis Bacon, Including all his Occasional Works, ed. James Spedding (7 vols.; London, 1861–74).
Letters and State Papers during the Reign of King James VI, ed. J. Maidment (Abbotsford Club; Edinburgh, 1838).
Letters of Queen Elizabeth and King James VI of Scotland, ed. John Bruce (Camden Society, 46; London, 1849).
[Lockhart, George], *Memoirs Concerning the Affairs of Scotland from Queen Anne's Accession to the Throne to the Commencement of the Union of the Two Kingdoms of Scotland and England*, 3rd edn. (London, 1714).
The Memoirs of Robert Carey, ed. F. H. Mares (Oxford, 1972).
Notes of the Treaty Carried on at Ripon between King Charles I and the Covenanters of Scotland, A.D. 1640, ed. John Bruce (Camden Society, 100; London 1869).
Original Letters Relating to the Ecclesiastical Affairs of Scotland, ed. D. Laing (2 vols.; Bannatyne Club; Edinburgh 1851).
The Parliamentary Diary of Robert Bowyer, 1606–1607, ed. David H. Willson (Minneapolis, Minn., 1931).
Proceedings in Parliament, 1610, ed. Elizabeth R. Foster (2 vols.; New Haven, Conn., 1966).
Rushworth, John, *Historical Collections* (8 vols.; London, 1721).
Secret History of the Court of James the First (2 vols.; Edinburgh, 1811).
Spain and the Jacobean Catholics, Vol. i: *1603–1612*, ed. Albert J. Loomie, SJ (Catholic Record Society, 64; London, 1973).

State Papers and Letters Addressed to William Carstares, ed. Joseph McCormick (Edinburgh, 1774).

Whitelocke, Sir Bulstrode, *Memorials of the English Affairs . . . to the End of the Reign of James I* (London, 1709..

Wilson, Arthur, *The History of Great Britain, being the Life and Reign of King James the First* (London, 1653).

Winwood, Sir Ralph, *Memorials of Affairs of State in the Reigns of Q. Elizabeth and K. James I*, ed. E. Sawyer (3 vols.; London, 1725).

The Writings and Speeches of Oliver Cromwell, ed. W. C. Abbott (4 vols.; Cambridge, Mass., 1937–47).

c. Tracts, Treatises, and Sermons

An Account of the Burning of the Articles of the Union at Dumfries (n.p., 1706).

The Advantages of Scotland by an Incorporate Union with England (n.p., 1706).

Advertisments of a Loyal Subject to his Gracious Soveraign, in *Somers Tracts*, ii (London, 1810), 144–8.

Anderson, James, *An Historical Essay Showing that the Crown of the Kingdom of Scotland is Imperial and Independent*, (Edinburgh, 1705).

The Antiquity of Englands Superiority over Scotland and the Equity of Incorporating Scotland, or other Conquered Nation, into the Commonwealth of England (London, 1652).

Atwood, William, *The Fundamental Constitution of the English Government* (London, 1690).

——, *The Superiority and Direct Dominion of the Imperial Crown of England, Over the Crown and Kingdom of Scotland* (London, 1704).

——, *The Scotch Patriot Unmask'd* (London, 1705).

Bacon, Sir Francis, *A Brief Discourse touching the Happy Union of the Kingdoms of England and Scotland* (London, 1603); reprinted in *The Life and Letters of Francis Bacon*, ed. James Spedding (London, 1864), iii. 89–99.

——, 'Certain Articles, or Considerations touching the Union of the Kingdoms of England and Scotland', in *Life and Letters of Francis Bacon*, iii. 218–34.

——, 'A Preparation toward the Union of the Laws of England and Scotland', in *The Works of Francis Bacon*, ed. Basil Montague, (3 vols.; Philadelphia, Pa., 1855).

Balfour, Sir James, *The Practicks of Sir James Balfour of Pittendreich*, ed. P. G. B. McNeill (2 vols.; Stair Society, 21–2; Edinburgh 1962–3).

Bates, Issac, *Two (United) are Better than One Alone* (London, 1707).

Bella Scot-Anglica: A Brief of all the Battells and Martiall Encounters which have happened 'twixt England and Scotland (London, 1648).

[Black, William], *Essay upon Industry and Trade* (Edinburgh, 1706)
——, *A Letter Concerning the Remarks upon the Considerations of Trade* [n.p., 1706].
——, *A Short View of Our Present Trade and Taxes* [Edinburgh, 1706]
——, *Some Considerations in Relation to Trade*, (Edinburgh, 1706).
A Breviate of the State of Scotland (London, 1689).
A Brotherly Exhortation from the General Assembly of the Church of Scotland to their Brethren of England (Edinburgh, 1649).
[Bruce, Alexander], *A Discourse of a Cavalier Gentleman on the Divine and Humane Laws with Respect to the Succession* (n.p., 1706).
[Buchan, John, of Cairnbulg], *A Memorial Pointing to Some of the Advantages of the Union of the Two Kingdoms* (London, 1702).
[Calderwood, David], *The Altar of Damascus or the Pattern of the English Hierarchie and Church Policie obtruded upon the Church of Scotland* ([Amsterdam], 1621)
[——], *Quaeres concerning the State of the Church of Scotland* (n.p., 1638).
——, *The History of the Kirk of Scotland*, ed. Thomas Thomson (8 vols.; Wodrow Society; Edinburgh, 1842–9).
Campion, Thomas, *The Discription of a Maske . . . in Honour of the Lord Hayes and his Bride*, in *Campion's Works*, ed. Percival Vivian (Oxford, 1909), 57–76.
[Carstares, William], *The Scottish Toleration Argued: or, an Account of all the Laws about the Church of Scotland Ratify'd by the Union-Act* (London, 1712).
[Chamberlen, Hugh], *The Great Advantages to both Kingdoms of Scotland and England by an Union* (n.p., 1702).
Chapman, George, Ben Jonson and John Marston, *Eastward Ho*, ed. R. W. Van Fossen (Manchester, 1979).
[Clerk, John], *A Letter to a Friend, Giving an Account how the Treaty of Union has been Received Here* (Edinburgh, 1706).
Coke, Sir Edward, *The Fourth Part of the Institutes of the Laws of England* (London, 1644)
A Collection of Scarce and Valuable Tracts . . . Selected from . . . Libraries, particularly that of the Late Lord Somers (13 vols.; London, 1809–15).
A Convincing Reply to the Lord Beilhaven's Speech in Relation to the Pretended Independency of the Scottish Nation (London, 1706).
A Copie of a Letter from some of the Nobility of Scotland to King James VI (n.p., 1706).
[Cornwallis, Sir William], *The Miraculous and Happie Union of England and Scotland* (London, 1604).
Cowell, John, *Institutiones Juris Anglicani* (Cambridge, 1605).
——, *Institutes of the Lawes of England* (London, 1651).

[Cowper, William], *The Bishop of Galloway His Dikaiologie* (London, 1614).

Craig, Sir Thomas, *Scotland's Soveraignty Asserted*, (London, 1695).

——, *De Unione Regnorum Britanniae Tractatus*, ed. C. S. Terry (SHS, 60; Edinburgh, 1909).

——, *The Jus Feudale*, tr. James Avon Clyde (2 vols.; Edinburgh, 1934).

Daniel, Samuel, *The Complete Works in Prose and Verse of Samuel Daniel*, ed. A. B. Grosart (5 vols.; London, 1885).

The Declaration of the Commissioners for the Kingdom of Scotland (London, 1647).

A Declaration and Brotherly Exhortation of the General Assembly of the Church of Scotland to their Brethren of England (London, 1647).

A Declaration of the Committee of Estates of Scotland concerning their Proceedings in Opposition to the Late Unlawfull Engagement in England (Edinburgh, 1648).

A Declaration and Exhortation of the General Assembly of the Church of Scotland to their Brethren of England (London, 1648).

Defoe, Daniel, *An Essay at Removing National Prejudices against a Union with Scotland*, Pts. I and II (London, 1706).

——, *A Discourse upon an Union between the Two Kingdoms of England and Scotland* (London, 1707).

——, *Remarks upon the Lord Havarsham's Speech in the House of Peers, February 15, 1707* [Edinburgh, 1707].

——, *The True-born Britain* (London, 1707).

——, *Union and No Union* (London, 1713).

——, *The History of the Union between England and Scotland* (London, 1786).

Dent, Giles, *A Thanksgiving Sermon Preach'd on the First Day of May, 1707* (London, 1707).

A Discourse concerning the Union [Edinburgh, 1706]

'A Discourse upon the Union of England and Scotland' (1664), in *Miscellanea Aulica*, ed. Thomas Brown (London, 1702), 192–8.

The Dissenters in England Vindicated (n.p., 1707).

Doddridge, John, 'A Breif Consideracion of the Unyon of Twoe Kingedomes' (1604), in *Jacobean Union*, 142–59.

——, *The English Lawyer* (London, 1631).

Drake, James, *Historia Anglo-Scotica: or an Impartial History of all that Happen'd between the Kings and Kingdoms of England and Scotland from the Beginning of the Reign of William the Conqueror to the Reign of Queen Elizabeth* (London, 1703).

Drayton, Michael, *The Works of Michael Drayton*, ed. J. W. Hebel (Oxford, 1933).

Duck, Sir Arthur, *De Usu et Authoritate Juris Civilis Romanorum in Dominiis Principum Christianorum* (London, 1653).

Eliot, Sir John, *De Jure Majestate*, ed. A. B. Grosart (London, 1882).

An Essay upon the Union of the Kingdoms of England and Scotland (London, 1705), in *Somers Tracts* xii (1814), 510–19.

[Fairfax, Blackerby], *A Discourse upon the Uniting Scotland with England* (London, 1702).

[Fletcher, Andrew of Saltoun], *State of the Controversy betwixt United and Separate Parliaments* (n.p., 1706).

Fortescue, Sir John, *The Governance of England*, ed. Charles Plummer (Oxford, 1885).

——, *De Laudibus Legum Anglie*, ed. S. B. Chrimes (Cambridge, 1942).

Gentili Alberico, 'De Unione Regnorum Britanniae', in *Regales Disputationes Tres* (London, 1605).

Gordon, John, *A Panegyrique of Congratulation for the Concord of the Realmes of Great Britaine in Unitie of Religion* (London, 1603).

——, EnΩtikon, or a Sermon of the Union of Great Brittannie in Antiquitie of Language, Name, Religion and Kingdome (London, 1604).

[Grant, Francis], *The Patriot Resolved in a Letter to an Addresser* (n.p., 1707).

The Grounds of the Present Danger of the Church of Scotland, [Edinburgh, 1707].

Hale, Sir Matthew, *The History of the Common Law of England*, ed. Charles M. Gray (Chicago, 1971).

Hamilton, John, 2nd Lord Belhaven, *A Speech in Parliament on the 10 Day of January 1701 by the Lord Belhaven* (Edinburgh, 1701).

——, *The Lord Belhaven's Speech in Parliament the second day of November 1706* (n.p., 1706).

Harington, Sir John, *A Tract on the the Succession to the Crown (A.D. 1602)*, ed. Clements R. Markham (Roxburghe Club; London, 1880).

The Harleian Miscellany, ed. T. Park (10 vols.; London, 1809–13).

Harrison, Stephen, *The Arches of Triumph Erected in Honour of the High and Mighty Prince James* (London, 1604).

Hayward, John, *A Treatise of Union of the Two Realmes of England and Scotland* (London, 1604).

Heylyn, Peter, *Aerius Redivivus* (London, 1672).

Hibernica, or Some Antient Pieces relating to Ireland, ed. Walter Harris, 2 pts. (Dublin, 1747–50).

The History of the Union of the Four Famous Kingdoms of England, Wales, Scotland and Ireland (London, 1660).

[Hodges, James], *The Rights and Interests of the Two British Monarchies, Treatise I* (London, 1703).

——, *The Rights and Interests of the Two British Monarchies, Treatise III* (London, 1706).

——, *Essay upon the Union* (Edinburgh, 1706).

[Howel, James] *The Grounds and Reasons of Monarchy Considered out of the Scottish History* (Edinburgh, 1651).

The Humble Address of the Presbytrie of Lanerk [n.p., 1706].

The Humble Address of the Presbytry of Hamilton (Edinburgh, 1706).

Hume, David, *De Unione Insulae Britanniae Tractatus 1* (London, 1605).

[Humfrey, John], *A Draught for a National Church Accommodation* (London, 1705).

Information from the Estaits of the Kingdome of Scotland to the Kingdome of England (n.p., 1640).

The Jacobean Union: Six Tracts of 1604, ed. Bruce Galloway and Brian P. Levack (SHS; Edinburgh, 1985).

James VI and I, *The Political Works of James I*, ed. Charles Howard McIlwain (Cambridge, Mass., 1918).

[Kennett, White], *A Letter from the Borders of Scotland*, (London, 1702).

Knox, John, *The Works of John Knox*, ed. David Laing (6 vols.; Edinburgh, 1846–64).

A Letter to a Friend concerning a French Invasion (London, 1692).

Lloyd, L. *The Jubile of Britane* (London, 1607).

Mackenzie, Sir George of Rosehaugh, *The Institutions of the Law of Scotland* (Edinburgh, 1684).

——, *A Vindication of the Government in Scotland during the Reign of King Charles II (Edinburgh,* 1691).

——, *Memoirs of the Affairs of Scotland, ed. T. Thomson (Edinburgh,* 1821).

[Mackenzie, George, Earl of Cromarty], *Parainesis Pacifica: or a Persuasive to the Union of Britain* (Edinburgh, 1702).

——, *A Letter to a Member of Parliament upon the 19th Article of the Treaty of Union between the two Kingdoms of Scotland and England* (n.p., 1706).

——, *Two Letters Concerning the Present Union, from a Peer in Scotland to a Peer in England* (Edinburgh, 1706).

Monson, Sir William, 'A Discovery of the Hollanders' Trades . . .', in J. Churchill, *A Collection of Voyages and Travels* (London, 1732), iii. 465–500.

Mosse, Miles, *Scotlands Welcome* (London, 1603).

[Mudie, Alexander], *Scotiae Indiculum* (London, 1682).

[Nicolson, William (ed.)], *Leges Marchiarum, or Border-Laws* (London, 1705).

Ollyffe, John, *A Sermon Preach'd May the 4th 1707* (London, 1707).

Overture for an Additional Clause to the Nineteenth Article, anent the Session in Scotland [n.p., 1706].

[Parsons, Robert], *A Conference about the Next Succession to the Crowne of Ingland* (n.p., 1594).

[Paterson, William], *An Inquiry into the Reasonableness and Consequences of an Union with Scotland* (London, 1706).

Paxton, Peter, *A Scheme of Union between England and Scotland* (London, 1705).

Pead, Deuel, *The Honour, Happiness and Safety of Union* (London, 1707).

Petyt, Sir William, 'Britannia Languens', in *Early English Tracts on Commerce*, ed. J. P. McCulloch (Cambridge, 1952).

Pont, Robert, 'Of the Union of Britayne' (1604), in *Jacobean Union*, 1–38.

A Protestation and Testimony against the Incorporating Union with England [n.p., 1707].

[Prynne, William], *Scotlands Ancient Obligation to England* (London, 1646).

Pyle, Thomas, *National Union a National Blessing* (London, 1707).

Queries to the Presbyterian Noblemen and Gentlemen, Barons, Burgesses, Ministers and Commoners in Scotland who are for the Scheme of an Incorporating Union with England [Edinburgh, 1706].

Rapta Tatio: The Mirrour of His Majesties Present Government, tending to the Union of his whole Iland of Brittonie (London, 1604).

Reflections upon a Late Speech by the Lord Haversham in so far as it relates to the Affairs of Scotland (Edinburgh, 1704).

Remarks upon a Late Dangerous Pamphlet, intitled The Reducing of Scotland by Arms (London, 1705).

Ridpath, George, *A Discourse upon the Union of Scotland and England* (London, 1702).

——, *Considerations upon the Union of the two Kingdoms* (London, 1706).

——, *The Reducing of Scotland by Arms and Annexing it to England as a Province Considered* (London, n.d.).

The Right of Succession to the Crown and Sovereignty of Scotland Argued (London, 1705).

Russell, John, 'A Treatise of the Happie and Blissed Unioun' (1604), in *Jacobean Union*, 75–141.

[Rutherford, Samuel], *Lex Rex: the Law and Prince* (London, 1644).

[Sage, J.], *Some Remarks on the Late Letter from a Gentleman in the City* (n.p., 1703).

Saltern, George, *Of the antient Lawes of Great Britaine* (London, [1605]).

Savile, Sir Henry, 'Historicall Collections' (1604), in *Jacobean Union*, 184–240.

A Scheme for Uniting the Two kingdoms of England and Scotland [Edinburgh, 1706].

The Scots Commissioners, their Desires concerning Unitie in Religion (London, 1641).

The Scottish Mist Dispel'd: or, A Clear Reply to the Prevaricating Answer of the Commissioners of the Kingdom of Scotland (London, 1648).

Selden, John, *Mare Clausum: The Right and Dominion of the Sea in Two Books* (London, 1663).

[Seton, William of Pitmedden], *The Interest of Scotland in Three Essays* (2 edn., Edinburgh, 1702.

——, *Scotland's Great Advantages by a Union with England* (n.p., 1706), in *Somers Tracts*, xii. (1814), 519–24.

——, *A Speech in Parliament the second day of November 1706* ([Edinburgh], 1706).

[Settle, Elkanah], *Carmen Irenicum: the Union of the Imperial Crowns of Great Britain* (London, 1707).

A Short Declaration of the Kingdom of Scotland for Information and Satisfaction to their Brethren of England (Edinburgh, 1644).

A Short Relation of the State of the Kirk of Scotland Since the Reformation of Religion (Edinburgh, 1638).

The Smoaking Flax Unquenchable: where the Union betwixt the two Kingdoms is Dissecated, Anatomized, Confuted and Annuled (n.p., 1706).

Some Papers of the Commissioners of Scotland Given in Lately to the Houses of Parliament concerning the Proposition of Peace (London, 1646).

Speed, John, *Theatre of the Empire of Great Britaine* (London, 1611).

Spelman, Sir Henry, 'Of the Union' (1604), in *Jacobean Union*, 160–83.

[Spence, Thomas], *The Testamentary Duty of the Parliament of Scotland with a view to the Treaty of Union* (n.p., 1707).

[Spottiswoode, John], *The Trimmer: or, Some Necessary Cautions Concerning the Union of the Kingdoms of Scotland and England* (Edinburgh, 1706).

Spreull, John, *An Accompt Current betwixt Scotland and England Ballanced* (Edinburgh, 1705).

The State of Scotland under the Past and Present Administration (n.p., 1703).

[Strachan, William], *A Toleration in Scotland no Breach of the Union* (London, 1713).

[Symson, David], *Sir George M'Kenzie's Arguments against an Incorporating Union Particularly Considered* (Edinburgh, 1706).

Tatham, John, *The Scots Figgaries: or, A Knot of Knaves. A Comedy* (London, 1652).

Thompson, John, Baron Haversham, *The Lord Haversham's Speech in the House of Peers on Saturday, February 15, 1706–7* (London, 1707).

Thornborough, John, *A Discourse Plainely Proving the Evident Utilitie and Urgent Necessitie of the Desired Happie Union of the two famous Kingdomes of England and Scotland* (London, 1604).

——, *The Ioiefull and Blessed Reuniting the two Mighty and Famous Kingdomes, England and Scotland* (Oxford, [1604]).

'Tom Tell-Troath', in *Complaint and Reform in England, 1436–1714*, ed. William H. Dunham, Jr. and Stanley Pargellis (New York, 1938).

'A Treatise about the Union of England and Scotland', (1604), in *Jacobean Union*, 39–74.

A True Account of the Great Expressions of Love from the Noblemen, Ministers and Commons of the Kingdom of Scotland unto Lieutenant General Cromwell (London, 1648).

Urquhart, Sir Thomas, *Tracts of the Learned and Celebrated Antiquarian Sir Thomas Urquhart of Cromarty* (London, 1774).

A View from the North or an Answer to the Voice from the South [Edinburgh, 1707].

Vulpone: or, Remarks on Some Proceedings in Scotland Relating both to the Union and Protestant Succession since the Revolution (1707).

[Webster, James], *The Author of the Lawful Prejudices against an incorporating Union Defended* (Edinburgh, 1707).

——, *Lawful Prejudices against an Incorporating Union with England* (Edinburgh, 1707).

[Willet, Andrew], *Ecclesia Triumphans* (Cambridge, 1603).

Williams, Daniel, *A Thanksgiving-Sermon, Occasioned by the Union of England and Scotland* (London, 1707).

Wilson, Thomas, 'The State of England, Anno Dom. 1600', ed. F. J. Fisher, in *The Camden Miscellany*, vol. 16 (Camden Society, 3rd ser. 52; London, 1936).

[Wright, William], *The Comical History of the Marriage-Union betwixt Fergusia and Heptarchus* (n.p., 1706).

Wyllie, Robert, *A Speech without Doors, concerning Toleration* (n.p., 1703).

——, *A Letter concerning the Union, with Sir George Mackenzie's Observations and Sir John Nisbet's Opinion upon the Same Subject* (n.p., 1706).

Zouch, Richard, *Elementa Jurisprudentiae* (Oxford, 1636).

SECONDARY SOURCES

Akzin, Benjamin, *State and Nation* (London, 1964).

Ashton, Robert, *The English Civil War: Conservatism and Revolution, 1603–1649* (London, 1978).

Aylmer, G. E., *The King's Servants: The Civil Service of Charles I* (London 1961).

Bagwell, Richard, *Ireland under the Stuarts and during the Interregnum* (3 vols.; London, 1909).

Bailyn, Bernard, *The Ordeal of Thomas Hutchinson* (Cambridge, Mass., 1974).

Bauckham, Richard, *Tudor Apocalypse* (Abingdon, 1978).

Bindoff, S. T., 'The Stuarts and their Style', *English Historical Review*, 60 (1945), 192–216.

Bogdanor, Vernon, *Devolution* (Oxford, 1979).

Bolton, G. C., *The Passing of the Irish Act of Union* (Oxford, 1966).

Boyd, J. P., *Anglo-American Union: Joseph Galloway's Plan to Preserve the British Empire, 1774–88* (Philadelphia, Pa., 1941).

Brown, P. Hume (ed.), *the Union of 1707: A Survey of Events* (Glasgow, 1907).

——, *The Legislative Union of England and Scotland* (Oxford, 1914).

Bruce, John, *Report on the Events and Circumstances which Produced the Union of England and Scotland* (London, 1799).

Burgess, R., *Perpetuities in Scots Law* (Stair Society, 31; Edinburgh 1979).

Burleigh, J. H. S., *A Church History of Scotland* (London, 1960).

Burrell, Sidney A., 'The Covenant Idea as a Revolutionary Symbol: Scotland, 1596–1637', *Church History*, 27 (1958), 338–50.

Cairns, John W., 'Institutional Writings in Scotland Reconsidered', in *New Perspectives in Scottish Legal History*, ed. Albert Kiralfy and Hector L. MacQueen (London, 1984), 76–117.

Calder, Angus, *Revolutionary Empire* (New York, 1981).

Cameron, J. T., 'Custom as a Source of Law in Scotland', *Modern Law Review*, 27 (1964), 306–21.

Carstairs, A. M., 'Some Economic Aspects of the Union of Parliaments', *Scottish Journal of Political Economy*, 2 (1955), 64–72.

Carswell, John, *From Revolution to Revolution: England 1688–1776* (New York, 1973).

Clarke, T. E. S. and H. C. Foxcroft, *A Life of Gilbert Burnet, Bishop of Salisbury* (London, 1907).

Cobbett, William, *The Parliamentary History of England from the Normans . . . to the Year 1803*, i (London, 1806).

Cochran-Patrick, R. W., *Records of the Coinage of Scotland from the Earliest period to the Union* (2 vols.; Edinburgh, 1876).

Collinson, Patrick, *The Elizabethan Puritan Movement* (London, 1967).

Coquillette, Daniel, 'Legal Ideology and Incorporation I: The English Civilian Writers, 1523–1607', *Boston University Law Review*, 61 (1980), 1–89.

Corrigan, Philip and Derek Sayer, *The Great Arch: English State Formation as Cultural Revolution* (Oxford, 1985).

Cowan, Edward J., 'The Union of the Crowns and the Crisis of the Constitution in 17th-Century Scotland', in *The Satellite State in the 17th and 18th Centuries*, ed. S. Dyrvik *et al.* (Bergen, 1979), 121–40.

Cowan, Ian B., *The Scottish Covenanters, 1660–1688* (London, 1976).

Daiches, David, *The Paradox of Scottish Culture: The Eighteenth-Century Experience* (London, 1964).

——, *Scotland and the Union* (London, 1977).

Davies, R. R., 'The Twilight of Welsh Law, 1284–1536', *History*, 56 (1966), 143–64.

Deutch, Karl W., *Nationalism and Social Communication*, (Cambridge, 1966).

——, and William T. Foltz (eds.), *Nation-Building* (New York, 1966).

Dicey, Albert V. and Robert S. Rait, *Thoughts on the Union between England and Scotland* (London, 1920).

Ditchfield, G. M., 'The Scottish Campaign against the Test Act, 1790–1791', *Historical Journal*, 23 (1980), 37–61.

Dodd, A. H., ' "A Commendacion of Welshmen" ', *Bulletin of the Board of Celtic Studies*, 19 (1962), 235–49.

Donahue, Charles, Jr., 'The Civil Law in England', *Yale Law Journal*, 84 (1974), 167–81.

Donaldson, Gordon, *The Making of the Scottish Prayer Book of 1637* (Edinburgh, 1954).

——, 'Foundations of Anglo-Scottish Union', in *Elizabethan Government and Society: Essays presented to Sir John Neale*, ed. S. T. Bindoff *et al.* (London, 1961), 282–314.

——, *Scotland: James V to James VII* (Edinburgh, 1971).

Dow, F. D., *Cromwellian Scotland, 1651–1660* (Edinburgh, 1979).

Drummond, Andrew L. and James Bulloch, *The Scottish Church, 1688–1843: The Age of the Moderates* (Edinburgh, 1973).

Duncan, A. M., *The Nation of the Scots and the Declaration of Arbroath* (London, 1970).

Dyson, Kenneth H. F., *The State Tradition in Western Europe: The Study of an Idea and an Institution* (Oxford, 1980).

Earle, Peter, *The World of Defoe* (London, 1976).

East, W. G., *The Union of Moldavia and Wallachia, 1859: An Episode in Diplomatic History* (Cambridge, 1929).

Eccleshall, Robert, *Order and Reason in Politics* (Oxford, 1978).

Elliott, J. H., 'The King and the Catalans, 1621–1640', *Cambridge Historical Journal*, 11 (1955), 253–72.

——, *Imperial Spain, 1469–1716* (Harmondsworth, 1970).

Enright, Michael J., 'King James and his Island: An Archaic Kingship Belief?', *SHR* 55 (1976), 29–40.

Epstein, Joel J., 'Sir Francis Bacon and the Issue of Union, 1603–1608', *Huntington Library Quarterly*, 33 (1970), 121–32.

Everitt, Alan, *The Community of Kent and the Great Rebellion, 1640–1660* (Leicester, 1973).

Feenstra, R. and C. J. D. Waal, *Seventeenth-Century Leyden Law Professors and their Influence on the Development of the Civil Law* (Amsterdam and Oxford, 1975).

Ferguson, William, 'The Making of the Treaty of Union of 1707', *SHR* 43 (1964), 89–110.

——, 'Imperial Crowns: A Neglected Facet of the Background to the Treaty of Union of 1707', *SHR* 53 (1974), 22–44.

——, *Scotland's Relations with England: A Survey to 1707* (Edinburgh, 1977).

Finer, Samuel E., 'State and Nation-Building in Europe: The Role of the Military', in *The Formation of National States in Western Europe*, ed. Charles Tilly (Princeton, NJ, 1975), 84–163.

Firth, C. H., 'Ballads Illustrating the Relations of England and Scotland during the Seventeenth Century', *SHR* 6 (1909), 113–28.

——, *The Last Years of the Protectorate, 1656–1658* (London, 1909).

——, 'The Ballad History of the Reign of James I', *TRHS*, 3rd ser. 5 (1911), 21–61.

——, 'The British Empire', *SHR* 15 (1918), 185–9.

Firth, Katharine R., *The Apocalyptic Tradition in Reformation Britain, 1530–1645* (Oxford, 1979).

Foster, Walter R., *Bishop and Presbytery: The Church of Scotland, 1661–1688* (London, 1958).

——, *The Church before the Convenants: The Church of Scotland, 1596–1638*, (Edinburgh, 1975).

Fraser, George MacDonald, *The Steel Bonnets* (New York, 1972).

Fulton, T. W., *The Sovereignty of the Seas* (Edinburgh, 1911).

Gay, Lance C., 'The Border Commissions: Law, Politics and Faction in the Stuart Middle Shires', MA thesis (Univ. of Maryland, 1972).

Gibb, Andrew D., *Law from over the Border: A Short Account of a Strange Jurisdiction* (Edinburgh, 1950).

Gordon, D. J., *The Renaissance Imagination*, ed. Stephen Orgel (Berkeley, Calif., 1975).

Gough, J. W., *Fundamental Law in English History* (Oxford, 1955).

Graham, Ian C. C., *Colonists from Scotland: Emigration to North America, 1707–1783* (Ithaca, NY, 1956).

Graham, John, *The Condition of the Border at the Union: The Destruction of the Graham Clan*, 2nd edn. (Glasgow, 1907).

Green, I. M., *The Re-establishment of the Church of England, 1660–1663* (Oxford, 1978).

Gunn, J. A. W., 'The Civil Polity of Peter Paxton', *Past & Present*, 40 (1968), 42–57.

Hall, Basil, 'Daniel Defoe and Scotland', in *Reformation, Conformity and Dissent*, ed. R. Buick Knox (London, 1977), 221–39.

Haller, William, *The Elect Nation* (New York, 1963).

Ham, R. E., *The County and the Kingdom: Sir Herbert Croft and the Elizabethan State* (Washington, DC, 1977).

Hamilton, Charles. L., 'The Basis for Scottish Efforts to Create a Reformed Church in England, 1640–1', *Church History*, 30 (1961), 171–8.

——, 'The Anglo-Scottish Negotiations of 1640–1', *SHR* 41 (1962), 84–6.

Hanham, H. J., *Scottish Nationalism* (London, 1969).

Hanson, Donald W., *From Kingdom to Commonwealth: The Development of Civic Consciousness in English Political Thought* (Cambridge, Mass., 1970).

Harvie, Christopher, *Scotland and Nationalism: Scottish Society and Politics, 1707–1977* (London, 1977).

Hay, Denys, 'The Use of the Term "Great Britain" in the Middle Ages', *Proceedings of the Society of Antiquaries of Scotland*, 81 (1958), 55–66.

——, 'England, Scotland and Europe: the Problem of the Frontier', *TRHS*, 5th ser. 25 (1975), 77–91.

Hechter, Michael, *Internal Colonialism: The Celtic Fringe in British National Development, 1536–1966*, (London, 1975).

Henderson, G. D., *Religious Life in Seventeenth-Century Scotland* (Cambridge, 1937).

Hetherington, W. M., *History of the Church of Scotland* (New York, 1859).

——, *History of the Westminster Assembly of Divines*, 4th edn., ed. R. Williamson (Edinburgh, 1878).

Hilgarth, J. N. H., *The Spanish Kingdoms, 1250–1516*, Vol. ii: *1410–1516* (Oxford, 1978).

Hill, Christopher, *God's Englishman: Oliver Cromwell and the English Revolution* (London, 1970).

——, *The World Turned Upside Down* (London, 1972).

Hirst, Derek, *Representative of the People?: Voters and Voting under the Early Stuarts* (Cambridge, 1975).

Holdsworth, W. S., *A History of English Law* (16 vols; London, 1922–52).

Hughes, Edward, 'Negotiations for a Commercial Union between England and Scotland in 1688', *SHR* 24 (1927), 30–47.

Huizinga, Johan, 'Patriotism and Nationalism in European History', in *Men and Ideas: History, the Middle Ages, the Renaissance* (New York, 1959), 97–155.

Hutton, G. M., 'Stair's Public Career', in *Stair Tercentenary Studies*, ed. David M. Walker (Stair Society, 33; Edinburgh, 1981), 1–68.

Hutton, Ronald, *The Restoration: A Political and Religious History of England and Wales, 1658–1667* (Oxford, 1985).

Insh, George P., *Scottish Colonial Schemes* (Glasgow, 1922).

——, *The Darien Scheme* (London, 1947).

Johnston, S. H. F., 'The Scots Army in the Reign of Queen Anne', *TRHS*, 5th ser. 3 (1953), 1–21.

Jones, Sir Thomas A., *The Union of England and Wales* (London, 1937).

Jones, W. J., 'The Exchequer of Chester in the Last Years of Elizabeth I', in *Tudor Men and Institutions*, ed. Arthur J. Slavin (Baton Rouge, La., 1972), 123–170.

Judson, Margaret, *The Crisis of the Constitution* (New Brunswick, NJ, 1949).

——, *From Tradition to Political Reality: A Study of the Ideas Set Forth in Support of the Commonwealth and Government of England* (Hamden, Conn., 1980).

Kantorowicz, Ernst H., *The King's Two Bodies: A Study in Mediaeval Political Theology* (Princeton, NJ, 1957).

Kaplan, Lawrence, *Politics and Religion during the English Revolution: The Scots and the Long Parliament* (New York, 1976).

Keith, Theodora, 'The Economic Condition of Scotland under the Commonwealth and Protectorate', *SHR* 5 (1908), 273–84.

——, *Commercial Relations of England and Scotland, 1603–1707* (Cambridge, 1910).

Kenyon, J. P., *Stuart England* (Harmondsworth, 1978).

Kettner, James H., *The Development of American Citizenship, 1608–1870* (Chapel Hill, NC, 1978).

Kiernan, V. G., 'State and Nation in Western Europe', *Past & Present*, 31 (1965), 20–38.

Kirk, James, ' "The Politics of the Best Reformed Kirks": Scottish Achievements and English Aspirations in Church Government after the Reformation', *SHR* 59 (1980), 22–53.

Knafla, Louis A., *Law and Politics in Jacobean England: the Tracts of Lord Chancellor Ellesmere* (Cambridge, 1977).

Koenigsberger, H. G., 'Early Modern Revolutions', *Journal of Modern History*, 46 (1974), 99–106.

Kohn, Hans, *The Idea of Nationalism* (New York, 1954).

Larner, Christina, *Enemies of God: The Witch-Hunt in Scotland* (Baltimore, Md., 1981).

Lee, Maurice, Jr., 'Comment on the Restoration of the Scottish Episcopacy, 1660–1661', *Journal of British Studies*, 1 (1962), 52–3.

——, *The Cabal* (Urbana, Ill., 1965).

——, *James I and Henri IV: An Essay in English Foreign Policy* (Urbana, Ill., 1970).

——, *Government by Pen: Scotland under James VI and I* (Urbana, Ill., 1980).

——, *The Road to Revolution: Scotland under Charles I, 1625–1637* (Urbana, Ill., 1985).

Lenman, Bruce, *The Jacobite Clans of the Great Glen, 1650–1784* (London, 1984).

Levack, Brian P., *The Civil Lawyers in England, 1603–1641: A Political Study* (Oxford, 1973).

——, 'The Proposed Union of English Law and Scots Law in the Seventeenth Century', *JR* NS 20 (1975), 97–115.

——, 'English Law, Scots Law and the Union, 1603–1707', in *Law-Making and Law-Makers in British History*, ed. Alan Harding (London, 1980), 105–19.

——, 'Toward a More Perfect Union: England, Scotland and the Constitution', in *After the Reformation; Essays in Honor of J. H. Hexter*, ed. B. Halament (Philadelphia, Pa., 1980), 57–74.

Lewis, Anthony M., 'Jefferson's *Summary View* as a Chart of Political Union', *William and Mary Quarterly*, 3rd ser. 5 (1948), 34–51.

Lindgren, R. E., *Norway–Sweden: Union, Disunion, and Scandinavian Integration* (Princeton, NJ, 1959).

Lloyd, Howell A., *The State, France and the Sixteenth Century* (London, 1983).

Loomie, Albert J., SJ, 'Philip III and the Stuart Succession in England', *Revue Belge de Philologie et d'Histoire*, 43 (1965), 492–514.

Lovat-Fraser, J. A., 'The Constitutional Position of the Scottish Monarch prior to the Union', *Law Quarterly Review*, 17 (1901), 252–62.

Luig, Klaus, 'The Institutes of National Law in the Seventeenth and Eighteenth Centuries', *JR*, NS 17 (1972), 193–226.

Lyall, Francis, *Of Presbyters and Kings: Church and State in the Law of Scotland* (Aberdeen, 1980).

Lythe, S. G. E., 'The Union of the Crowns and the Debate on Economic Integration', *Scottish Journal of Political Economy* 5 (1958), 219–28.

MacCormick, Neil (ed.), *The Scottish Debate: Essays on Scottish Nationalism* (London, 1970).

McIlwain, Charles Howard, *The American Revolution: A Constitutional Interpretation* (New York, 1924).

McKechnie, W. S., 'The Constitutional Necessity for the Union of 1707', *SHR* 5 (1908), 52–66.

Mackenzie, W. J. M., 'Peripheries and Nation-Building: the Case of Scotland', in *Mobilization, Center-Periphery Structures and Nation-Building*, ed. Peter Torsvik (Oslo, 1981), 153–80.

Mackinnon, James, *The Union of England and Scotland* (London, 1896).

MacLean, A. J., 'The 1707 Union: Scots Law and the House of Lords', in *New Perspectives in Scottish Legal History*, ed. Albert Kiralfy and Hector L. MacQueen (London, 1984), 50–75.

McLeod, W. R. and V. B., *Anglo-Scottish Tracts, 1701–1714: A Descriptive Checklist* (Lawrence, Kan., 1979).

McMahon, George I. R., 'The Scottish Courts of High Commission, 1610–38', *Records of the Scottish Church History Society*, 15. 3 (1965), 193–209.

McMillan, A. R. G., 'The Judicial System of the Commonwealth of Scotland', *JR* 49 (1937), 232–55.

——, *The Evolution of the Scottish Judiciary* (Edinburgh, 1941).

MacNair, Alfred Duncan, Lord, *The Law of Treaties* (Oxford, 1961).

Maitland, F. W., 'The Crown as Corporation', *Law Quarterly Review*, 17 (1901), 131–346.

Makey, Walter, *The Church of the Covenant, 1637–1651: Revolution and Social Change in Scotland* (Edinburgh, 1979).

Mann, Arthur, *The One and the Many: Reflections on the American Identity* (Chicago, 1979).

Manning, Roger B., *Religion and Society in Elizabethan Sussex* (Leicester, 1969).

Maravall, José Antonio, 'The Origins of the Modern State', *Journal of World History*, 6 (1960–1), 788–808.

Marshall, Ebenezer, *The History of the Union of Scotland and England* (Edinburgh, 1799).

Mathieson, William L., *Politics and Religion: A Study in Scottish History from the Reformation to the Revolution* (2 vols.; Glasgow, 1902).

——, *Scotland and the Union* (Glasgow, 1905).

Matthews, Nancy L., *William Sheppard: Cromwell's Law Reformer* (Cambridge, 1984).

Millard, A. M., 'The Import Trade of London, 1600–1640', Ph.D. thesis (London, 1956).

Mitchison, Rosalind, 'Patriotism and National Identity in Eighteenth-Century Scotland', in *Nationality and the Pursuit of National Independence*, ed. T. W. Moody (Belfast, 1978), 73–95.

——, *Lordship to Patronage: Scotland, 1603–1745* (London, 1983).

Mitteis, Heinrich, *The State in the Middle Ages: A Comparative Constitutional History of Feudal Europe*, tr. H. F. Orton (Amsterdam, 1975).

Moir, Thomas L., *The Addled Parliament of 1614* (Oxford, 1958).

Murdoch, Alexander, 'The Advocates, the Law and the Nation in Early Modern Scotland', in *Lawyers in Early Modern Europe and America*, ed. Wilfrid Prest (London, 1981), 147–63.

Murray, Andrew Graham, 1st Viscount Dunedin, *The Divergencies and Convergencies of English and Scottish Law* (Glasgow, 1935).

Murray, Athol L., 'Administration and the Law', in *The Union of 1707: Its Impact on Scotland*, ed. Thomas I. Rae (Glasgow, 1974), 30–57.

Murray, Ronald K., 'The Anglo-Scottish Union', *Scots Law Times*, 4 Nov. 1961, 161–4.

——, 'Devolution in the U.K.: A Scottish Perspective', *Law Quarterly Review*, 96 (1980), 35–50.

Nevo, Ruth, *The Dial of Virtue: Poems on Affairs of State in the Seventeenth Century* (Princeton, NJ, 1963).

Newton, Robert, 'The Decay of the Borders: Tudor Northumberland in Transition', in *Rural Change and Urban Growth, 1500–1800: Essays in English Regional History in Honour of W. G. Hoskins*, ed. C. W. Chalkin and M. A. Havinden (London, 1974).

Nichols, John, *The Progresses, Processions and Magnificent Festivities of King James the First* (4 vols.; London, 1828).

Nobbs, D., *England and Scotland, 1560–1707* (London, 1952).

Notestein, Wallace, 'The Establishment of the Committee of Both Kingdoms', *AHR* 17 (1912), 477–95.

——, *The House of Commons, 1604–1610* (New Haven, Conn., 1971).

O'Day, Rosemary, *The English Clergy: The Emergence and the Consolidation of a Profession* (London, 1979).

Ogilvie, J. D., 'Church Union in 1641', *Records of the Scottish Church History Society*, 1 (1926), 143–60.

Omond, G. W. T., *The Early History of the Scottish Union Question* (Edinburgh, 1897).

Palliser, D. M., *The Age of Elizabeth: England under the Later Tudors, 1547–1603* (London, 1983).

Pantin, W. A., *Oxford Life in Oxford Archives* (Oxford, 1972).

Pares, Richard, 'A Quarter of a Millenium of Anglo-Scottish Union', *History*, 39 (1954), 233–8.

Parry, Graham, *The Golden Age Restor'd: The Culture of the Stuart Court, 1603–1642* (Manchester, 1981).

Paul, Robert S., *The Assembly of the Lord: Politics and Religion in the Westminster Assembly and the 'Grand Debate'* (Edinburgh, 1985).

Pawlisch, Hans S., *Sir John Davies and the Conquest of Ireland: A Study in Legal Imperialism* (Cambridge, 1985).

Peck, Linda Levy, *Northampton: Patronage and Policy at the Court of James I* (London, 1982).

Phillipson, N. T., 'The Scottish Whigs and the Reform of the Court of Session 1785–1830', Ph.D. thesis (Cambridge, 1967).

——, 'Nationalism and Ideology', in *Government and Nationalism in Scotland*, ed. J. N. Wolfe (Edinburgh, 1969).

——, 'Scottish Public Opinion and the Union in the Age of the Association' in *Scotland in the Age of Improvement*, ed. N. T. Phillipson and Rosalind Mitchison, (Edinburgh, 1970), 125–47.

——, 'Lawyers, Landowners, and the Civic Leadership of Post-Union Scotland', *JR*, NS 21 (1974), 97–120.

Pocock, J. G. A., *The Ancient Constitution and the Feudal Law*, (Cambridge, 1957; 2nd edn., Cambridge, 1987).

——, 'British History; A Plea for a New Subject', *Journal of Modern History*, 47 (1975), 601–28.

——, *The Machiavellian Moment: Florentine Political Thought and the Atlantic Republican Tradition* (Princeton, NJ, 1975).

——, 'The Limits and Divisions of British History', *AHR* 87 (1982), 311–36.

Poggi, Gianfranco, *The Development of the Modern State: A Sociological Introduction* (Stanford, Calif., 1978).

Povolich, Charles A., 'Scottish Juridical Institutions and Practices under the Impact of the Union with England', Ph.D. thesis (Univ. of Southern California, 1953).

Prest, Wilfrid, *The Inns of Court under Elizabeth I and the Early Stuarts, 1590–1640* (London, 1972).

Pryde, George S., *The Treaty of Union of Scotland and England, 1707* (London, 1950).

Rabb, Theodore, 'Sir Edwin Sandys and the Parliament of 1604–10', *AHR* 69 (1976), 646–70.

Rae, Thomas I., *The Administration of the Scottish Frontier, 1513–1603* (Edinburgh, 1966).

Rait, Robert S., *An Outline of the Relations between England and Scotland (500–1707)* (London, 1901).

Rees, William, 'The Union of England and Wales', *Transactions of the Honourable Society of Cymmrodorion (Session 1937)* (1938), 27–100.

Reid, Alistair, 'Digging Up Scotland', *The New Yorker*, 5 Oct. 1981, 59–125.

Riley, P. W. J., *The Union of England and Scotland: A Study in Anglo-Scottish Politics of the Eighteenth Century* (Manchester, 1978).

Roberts, P. R., 'The "Acts of Union" and the Tudor Settlement of Wales', Ph.D. thesis, (Cambridge, 1966).

Rogers, Charles, 'Memoir and Poems of Sir Robert Aytoun', *TRHS* 1 (1872), 104–219.

——, *Memorials of the Earl of Stirling and of the House of Alexander* (2 vols.; Grampian Club; London, 1877).

Roots, Ivan, *The Great Rebellion, 1642–1660* (London, 1966).

Rose, Richard, 'The United Kingdom as a Multi-National State', Survey Research Centre, Univ. of Strathclyde, Occasional Paper, 6 (Glasgow, 1970).

Russell, Elizabeth, 'The Influx of Commoners into the University of Oxford before 1581: An Optical Illusion?', *English Historical Review*, 92 (1977), 721–45.

Scarman, Sir Leslie, *English Law: the New Dimension* (London, 1974).

Scott, W. R., 'The Fiscal Policy of Scotland before the Union', *SHR* 1 (1904), 173–90.

Seddon, P. R., 'Patronage and Officers in the Reign of James I'. Ph.D. thesis (Manchester, 1967).

Sellar, W. David H., 'English Law as a Source', in *Stair Tercentenary Studies*, ed. David M. Walker (Stair Society, 33; Edinburgh, 1981), 140–50.

Seton-Watson, Hugh, *Nations and States* (Boulder, Colo., 1977).

Sharpe, Kevin, *Sir Robert Cotton, 1586–1631: History and Politics in Early Modern England* (Oxford, 1979).

Shennan, J. H., *The Origins of the Modern European State, 1450–1725* (London, 1974).

Simpson, A. W. B., 'Entails and Perpetuities', *JR* NS 24 (1979), 1–20.

Skinner, Quentin, 'History and Ideology in the English Revolution', *Historical Journal*, 8 (1965), 152–78.

Smith, D. B., 'Sir Thomas Craig, Feudalist', *SHR* 12 (1915), 271–302.

Smith, Janet Adam, 'Some Eighteenth-Century Ideas of Scotland', in *Scotland in the Age of Improvement*, ed. N. T. Phillipson and Rosalind Mitchinson (Edinburgh, 1970), 107–24.

Smith, T. B., *Studies Critical and Comparative* (Edinburgh, 1962).

——, 'Scottish Nationalism, Law and Self-Government', in *The*

Scottish Debate: Essays on Scottish Nationalism, ed. Neil MacCormick (London, 1970), 34–51.

——, 'Authors and Authority', *Journal of the Society of Public Teachers of Law*, NS 12 (1972), 3–21.

——, 'British Justice: A Jacobean Phantasma', *The Scots Law Times*, 4 June 1982, 157–64.

Smout, T. C., *Scottish Trade on the Eve of the Union, 1660–1707* (Edinburgh, 1963).

——, 'The Anglo-Scottish Union of 1707: I. The Economic Background', *Economic History Review*, 2nd ser. 16 (1964), 455–67.

——, 'The Road to Union', in *Britain After the Glorious Revolution*, ed. G. Holmes (London, 1969), 176–96.

——, 'Scotland in the Seventeenth Century: A Satellite Economy?', in *The Satellite State in the 17th and 18th Centuries*, ed. S. Dyrvik *et al.* (Bergen, 1979).

Snow, Vernon, *Essex the Rebel: The Life of Robert Devereux, the Third Earl of Essex, 1591–1646* (Lincoln, Nebr., 1970).

Stafford, Helen G., *James VI of Scotland and the Throne of England* (New York, 1940).

Stein, Peter, *Roman Law in Scotland*, Pt. 5, 13b of *Jus Romanum Medii Aevi* (Milan, 1968).

Stevenson, David, *The Scottish Revolution, 1637–1644: The Triumph of the Covenanters* (Newton Abbot, 1973).

——, 'The Radical Party in the Kirk, 1637–1645', *Journal of Ecclesiastical History*, 25 (1974), 135–65.

——, *Revolution and Counter-Revolution in Scotland, 1644–1651* (London, 1977).

Stone, Lawrence, 'State Control in Sixteenth-Century England', *Economic History Review*, 17 (1947), 103–20.

Strayer, Joseph R., *On the Medieval Origins of the Modern State* (Princeton, NJ, 1970).

Sutherland, Donald, 'Conquest and Law', *Studia Gratiana*, 15 (1972), 33–51.

Sutherland, Robert, 'Aspects of the Scottish Constitution Prior to 1707', in *Independence and Devolution: The Legal Implications for Scotland*, ed. John P. Grant (Edinburgh, 1976), 15–44.

Terry, C. S., *The Scottish Parliament: Its Constitution and Procedure, 1603–1607* (Glasgow, 1905).

Tilly, Charles (ed.), *The Formation of National States in Western Europe* (Princeton, NJ, 1975).

Tivey, Leonard (ed.), *The Nation-State: The Formation of Modern Politics* (Oxford, 1981).

Tough, Douglas L. W., *The Last Years of a Frontier* (Oxford, 1928).

Trevelyan, 'The Middle Marches', in *Clio, A Muse and Other Essays Literary and Pedestrian* (London, 1913).

Trevor-Roper, Hugh R., *Religion, the Reformation and Social Change* (London, 1967).

Tytler, Patrick F., *An Account of the Life and Writings of Sir Thomas Craig of Riccarton* (Edinburgh, 1823).

Ullmann, Walter, *Medieval Political Thought* (Harmondsworth, 1975).

Usher, Roland G., *The Rise and Fall of the High Commission* (Oxford, 1913).

Veall, Donald, *The Popular Movement for Law Reform, 1640–1660* (Oxford, 1970).

Venn, John and J. A., *The Book of Matriculations and Degrees, 1544–1659* (Cambridge, 1913).

Viner, Jacob, *The Customs Union Issue* (London, 1950).

Vives, J. Vincens, 'The Administrative Structure of the State', in *Government and Society in Reformation Europe, 1520–1560*, ed. H. J. Cohn (London, 1971), 58–87.

Wallace, John M., *Destiny His Choice: The Loyalism of Andrew Marvell* (Cambridge, 1968).

Wallerstein, Immanuel, *The Modern World-System: Capitalist Agriculture and the Growth of the European World Economy in the Sixteenth Century* (New York, 1976).

Walzer, Michael, *The Revolution of the Saints* (Cambridge, Mass., 1965).

Watson, Alan, *Legal Transplants: An Approach to Comparative Law* (Edinburgh, 1974).

——, *The Making of the Civil Law* (Cambridge, Mass., 1981).

Watts, Michael R., *The Dissenters* (Oxford, 1978).

Watts, S. J., *From Border to Middle Shire: Northumberland 1586–1625* (Leicester, 1975).

Webb, Stephen S., *The Governors-General: The English Army and the Definition of Empire* (Chapel Hill, NC, 1979).

Wedgwood, C. V., 'Anglo-Scottish Relations, 1603–1640', *TRHS*, 4th ser. 32 (1950), 31–48.

——, *Poetry and Politics under the Stuarts* (Cambridge, 1960).

Wheare, K. C., *Federal Government*, 4th edn. (London, 1963).

Wheeler, Harvey, 'Calvin's Case (1608) and the McIlwain-Schuyler Debate', *AHR* 61 (1956), 587–97.

White, Philip C., 'What is Nationality?', *Canadian Review of Studies in Nationalism*, 12 (1985), 1–23.

Williams, 'The Northern Borderland under the Early Stuarts', in *Historical Essays 1600–1750 Presented to David Ogg*, ed. H. E. Bell and R. L. Ollard (London, 1963), 1–17.

Williamson, Arthur H., *Scottish National Consciousness in the Age of James VI* (Edinburgh, 1979).

Willock, Ian D., *The Origins and Development of the Jury in Scotland* (Stair Society, 23; Edinburgh, 1966).

Willson, D. H., 'King James I and Anglo-Scottish Unity', in *Conflict in Stuart England: Essays in Honour of Wallace Notestein*, ed. William A. Aiken and Basil D. Henning (London, 1960), 43–55.

Wilson, Charles, *England's Apprenticeship, 1603–1763* (London, 1965).

Wilson, Nan, 'The Scottish Bar: The Evolution of the Faculty of Advocates in its Historical Social Setting', *Louisiana Law Review*, 28 (1968), 235–57.

——, 'The Professional Community: The Edinburgh Advocates', *The Record of the Association of the Bar of the City of New York*, 23 (1968), 174–84.

Wood, Anthony, *Fasti Oxoniensis* (2nd edn., 2 vols.; London, 1813–20).

Woodward, Douglas, 'Anglo-Scottish Trade and English Commercial Policy during the 1660s' *SHR* 56 (1977), 153–74.

Woolrych, Austin, 'Last Quests for a Settlement', in *The Interregnum*, ed. G. E. Aylmer (London, 1972).

Wormald, Jenny, *Court, Kirk and Community: Scotland 1470–1625* (London, 1981).

——, 'Gunpowder, Treason and Scots', *Journal of British Studies*, 24 (1985), 141–68.

Yates, Frances, *Astraea: The Imperial Theme in the Sixteenth Century* (London, 1975).

Zagorin, Perez, *Rebels and Rulers: 1500–1660* (2 vols.; Cambridge, 1981).

Index

Abbot, George 131
Aberdeen 125
acquisitive prescription, doctrine of 81
acts of parliament, British: civil jury trial (1815) 93; establishment of the High Court (1873) 95: for improving the Union (1708) 99; Toleration (1712) 134–5
acts of parliament, English 54, 92; in Restraint of Appeals (1533) 2; Union with Wales (1536 and 1543) 6, 18, 19; Alien (1705) 12, 220; Union with Scotland (1707) 12 n., 41 n., 102 n.; Security for Church of England (1706) 13, 102, 134; Bill of Rights (1689) 54; Limitations (1624) 81; *De Donis Conditionalibus* (1285) 92; *Quia Emptores* (1290) 92; Fines (1484 and 1489) 92; Uses (1536) 92; Wills (1540) 92; Toleration (1689) 133, 134; Navigation (1660) 142, 150, 151, 163, 167
acts of parliament, Scottish 33, 54, 70, 92; anent Peace and War (1703) 12, 218; Security (1703–4) 12, 218; Union with England (1707) 12 n., 41 n., 93, 102 n.; Security for Church of Scotland (1706) 13, 102, 134, 135; Union (1604) 46, 128; Claim of Right (1689) 54; anent the King's Prerogative (1606) 56; Supremacy (1669) 57, 131
Adam, James 206, 207
Adam, Robert 206, 207
administration, Scottish 62–6, 211
admiralty courts 81
adverse possession, doctrine of 81
Alford, Edward 39
American Revolution 183
Anderson, James 27
Anne, Queen of England and Scotland 11, 12, 13, 147, 215, 225
ante-nati 44, 182, 184, 187, 188
apocalyptic thought 107–9, 172–3, 176
appeals 98–9
Aragon, crown of 215 n.
Aragonese people 178
Aristotle 89
Armada, Spanish 21, 172

Arminianism 173
Ashe, Thomas 74 n.
Ashley, Sir Francis 80
Auchterlony, Sir William 187; son of 187
Auld Alliance 216

Bacon, Sir Francis: conflict of with Sandys 29; and union of laws 44, 74, 78, 84; and union of parliaments 45, 47; and the prerogative 56 n.; and single British council 61; and court at Berwick 62; and legal maxims 80; and civil law 80 n.; and religious union 103; and commercial union 139; and British history 212
Baillie, Robert 115
Balcanqual, Walter 186
Balfour, Arthur 209
Balfour, James, minister 119
Balfour, Sir James, of Pittendreich 83
Bancroft, Richard 124, 185
Bank of Scotland 167
Banqueting House, Whitehall 224
Bartolists 80
Belhaven, Lord *see* Hamilton
Bennet, Sir John 60, 81
Berwick 62, 71, 75, 158, 193 n.
Bethell, Sir Hugh 187; daughter of 187
bishops: English 114, 122, 125; Scottish 124, 125–6, 127, *see also* episcopacy
Bishops' War (1640) 9, 149, 198
Black, William 153–4
Board of Manufactures 65
body politic, theory of 7, 43, 69-70, *see also* king's two bodies
Bordeaux 158, 161
Border commissions 64, 113 n., 192
Border law 19
Border reivers 190
Borders, Borderlands 6, 8, 62, 63, 74, 82, 113 n., 180, 186, 190–3
Boswell, James 206
Bracton, Henri de 79
Brewster, Sir Francis 146
Brightman, Thomas 173
British Library 212
British Museum 212
Bruce, Alexander 33